EINSTEIN IN TIME AND SPACE

A Life in 99 Particles

SAMUEL GRAYDON

SCRIBNER

New York   London   Toronto   Sydney   New Delhi

Scribner
An Imprint of Simon & Schuster, Inc.
1230 Avenue of the Americas
New York, NY 10020

First Scribner hardcover edition November 2023

SCRIBNER and design are registered trademarks of The Gale Group, Inc., used under license by Simon & Schuster, Inc., the publisher of this work.

For information about special discounts for bulk purchases, please contact Simon & Schuster Special Sales at 1-866-506-1949 or business@simonandschuster.com.

The Simon & Schuster Speakers Bureau can bring authors to your live event. For more information or to book an event, contact the Simon & Schuster Speakers Bureau at 1-866-248-3049 or visit our website at www.simonspeakers.com.

Interior design by Jaime Putorti

Manufactured in the United States of America

10 9 8 7 6 5 4 3 2 1

Library of Congress Cataloging-in-Publication Data is available.

ISBN 978-1-9821-8510-7
ISBN 978-1-9821-8512-1 (ebook)

CONTENTS

INTRODUCTION

On May 29, 1919, above the skies of the island of Príncipe, the moon passed in front of the sun, and the world fell into shadow. The totality of the eclipse had begun. Waiting for just this moment, an English scientist peered through an astrographic camera to take pictures of the event. And in the Brazilian city of Sobral another scientist did the same, feverishly taking as many photographs as the few minutes of darkness would allow. They worked in the hope that they might record the bending of starlight. They succeeded.

When, a few months later, these results were announced at Burlington House in London, home of the Royal Society, the conventional understanding of gravity was swept away. The photographs captured by the expedition teams showed that when the light from stars 153 light-years away, at the center of the Taurus constellation, had reached the vicinity of the sun, it had altered its path, so that those stars no longer appeared in their usual positions in the sky. There was only one explanation that exactly accounted for this: space itself was warped by the presence of the sun. The theory of relativity had just been confirmed. Isaac Newton, the giant of physics, had been cast down, to be replaced by Albert Einstein, a scientist little known beyond Germany.

Einstein, then living in Berlin, forty years old with only the smallest trace of gray at his temples, received the results of the eclipse expedition shortly before he was due to meet with a student named Ilse Schneider. During their conversation, he handed her the telegram he'd received informing him of his theory's success. Recognizing the deep shift she must now make in her understanding of the universe and its laws, she was understandably excited and offered many congratulations, but Einstein told her calmly, "I *knew* the theory was correct."

But, she asked, what if those who observed the eclipse hadn't seen the bending of light? Or what if they'd seen the light bend, but not by the amount Einstein had predicted in his theory?

"Then I should feel very sorry for the Lord," he replied. "The theory is correct."

Two years later, Einstein toured America, to help raise money for the Zionist cause, which was focused on creating a Jewish home in Palestine. He was now as famous as it was possible to be. In every city he visited, thousands upon thousands of people filled the streets to see him. A crowd of admirers lifted him onto their shoulders, he met the president, and the Senate even undertook a discussion about the difficulty of understanding relativity. A year later, he was awarded the Nobel Prize in Physics and undertook a lecture tour of Asia. While in Japan, he met the emperor and empress, and a huge crowd waited all night outside his hotel, hoping to see the great man appear on his balcony. In Tokyo, he lectured for four hours with translation. Feeling a little sorry to have put an audience through such a slog, he managed to cut down his next talk on the tour to under three hours. As he rode the train to the next city, he noticed that his hosts were not quite themselves and asked if anything was wrong. Yes, came the reply: Einstein had insulted those who had organized the second lecture by making it shorter than the first. And so for the rest of the tour he made sure to take his time explaining himself, and the audiences were happy to listen.

Einstein's fame was so rapid and so complete that, around the same time, two American students, to play out a bet with one another, addressed an envelope to "Professor Albert Einstein, Europe" and posted it, wanting to see if it would reach him. It did, and with no more delay than usual.

"How excellent the postal service is!" was all Einstein said.

Twenty years before this, in 1902, Albert had just moved to the city of Bern in Switzerland. Twenty-three and a little round in the face, he emanated a restless energy, an easy intensity. A friend who met him for the first time in Bern was immediately struck by the "radiance of his large eyes." He was waiting to be offered a job at the Swiss Patent Office, in the city—thanks to the help of a friend, the job was effectively sewn up for him. Nevertheless, things were far from rosy. He had very little money and had resorted to advertising his services as a private tutor of physics and mathematics in the local paper. But students were few and his rates were low; he complained it would be easier to make a living busking in the street with his violin. He ate very little. Moreover, his girlfriend, Mileva Marić, had given birth to his daughter less than a month before his move. Were the existence of a child born outside marriage to be discovered, he could say goodbye to his post at the patent office. Albert and Mileva endeavored to keep her a secret from everyone, including Einstein's family. He knew he must marry—he wished to marry, he believed—but he had not yet gathered the courage to do so. His parents had long made clear their dislike for Mari, and he knew they certainly would not bless the union.

Added to this, although the position at the patent office was most welcome, in accepting it Einstein would be, in a way, accepting failure as well. In the two years since leaving university he had applied for academic positions all over Europe, only to be rejected every time. The patent office was a necessity, but it made definite Einstein's academic failing, his inability to pursue what he loved and the supremacy of obligations.

Einstein would continue to apply for academic posts for the next five years, before finally making it to the lowest rung on the ladder. At one point in his degrading job hunt, ground down by constant rejection, he modified his ambitions and applied for work as a high school teacher—sending copies of all his scientific papers, including his doctoral thesis and papers on light quanta and special relativity, as supporting documents. There were twenty-one applicants. Einstein didn't make the final three.

It is easy to think of Einstein's life as bifurcating into two halves: before and after the confirmation of general relativity, which is to say before and after fame. In his youth, according to this narrative, he was unappreciated and yet brilliant, while in his old age he was appreciated yet dull. There is some truth to this. Einstein's finest work was all produced before he was famous, and for much of his early life he was a reasonably obscure figure. It took him nine years to secure an assistant professorship, and even then he wasn't first choice for the job.

And it is true that after his fame he produced few papers worthy of note. What was perhaps Einstein's last really great piece of work was written twenty years before his death. And this was not completed in the trailblazing spirit of earlier work—it did not attempt to explain an unknown, say, or reshape any area of inquiry—but instead was reactionary, forged in distrust of the new physics of quantum mechanics. In it, Einstein set out to discredit the quantum by outlining "entanglement," a phenomenon that could theoretically occur under the rules of quantum mechanics, but that he thought was impossible in reality. One of Einstein's more remarkable habits was to be right even when he was wrong. In this case, entanglement was eventually proved to be one of the fundamental truths of the universe.

For most of the last thirty years of his life he would dedicate himself to developing a "unified field theory"—a theory of everything, under which all the laws of nature are encompassed, from the movements of celestial bodies to magnetism to what occurs inside the atom—only to be increasingly ignored by his scientific colleagues, who considered him something of a relic and unlikely ever to succeed.

And yet Einstein cannot be so simply defined and is far more interesting than this narrative of accolade harvesting and stagnation implies. It, too, neatly sands off the outlying facts, such as the professional recognition and success he had achieved in Germany even before publishing anything about general relativity. It also ignores his support for the Jewish people and the development of his pacifism. During the buildup to the Second World War these other aspects of the man were far from stagnant. He spent a large portion of his money helping Jews to flee Germany and immigrate to the US, as well as helping to found the organization that became the International Rescue Committee.

Einstein's fame can get in the way of an objective assessment of his life. It creates an expectation of the extraordinary, and so it's easy to fail to see what an astounding life Einstein did actually live. It involved a level of genuine success that is almost unthinkable. In one year—in fact, in half a year, from March to September 1905—he submitted his doctoral thesis; he mathematically proved the existence of atoms; he argued for the modern idea of light as a stream of particles (and in so doing laid the groundwork for quantum mechanics); and he proposed the special theory of relativity—in the process doing away with the past few hundred years of scientific orthodoxy and, practically by accident, discovering the equivalence of energy and matter, now immortalized in the equation $E = mc^2$. He did all this in his spare time, while working six days a week as a patent clerk, without access to a library, and with a one-year-old child at home.

On top of this, ten years later, he presented the general theory of relativity, which managed, in one set of equations that exhibited an incredible degree of accuracy, to set out the laws that govern the star-paved heavens. Almost alone, Einstein had figured out a way of conceiving of space that could describe exactly the movements of the objects in space: it could account for the orbit of Mercury, the motion of two stars orbiting each other, and a thousand more situations. General relativity was so successful in describing the workings of the universe that it anticipated truths even Einstein couldn't quite believe. Einstein thought the universe was static, but his theory demanded that it be expanding; the theory was right. Relativity insisted on the existence of strange objects in space that are so dense nothing can escape their gravity. Einstein thought these a mathematical bug that could be ignored. They turned out to be black holes, and very real indeed.

Not just early on but later in life, too, Einstein experienced hardship that was nearly as dramatic as his achievements. Einstein and Marić eventually gave up their daughter entirely, something that gravely affected their own relationship. Later, his acrimonious divorce led to a complicated, bitter, and sad relationship with his remaining two children, Hans Albert and Eduard. Things unraveled particularly with Eduard, who, at the age of twenty, threatened suicide and later underwent treatment for schizophrenia, spending much of his remaining life in and out of asylums. Einstein was twice a potential target for assassination, and after the ascension of the Nazi Party, in what was the most extreme expression of antisemitism he experienced during his entire life, he was made an exile from Germany, from his home, possessions, and friends.

* * *

But, for all that, Einstein was in many ways really quite normal—in him, the fabled idea that genius and madness are two aspects of the same state isn't borne out. He wasn't reclusive, but made friends effortlessly and navigated those relationships with ease. Far from being monomaniacal, he took an interest in music, art, and psychology and was a vocal and active participant in the politics of the day. At various times, he was a founder of the pacifist organization the New Fatherland League, served on the League of Nations' International Committee on Intellectual Cooperation, and was co-chair of the American Crusade to End Lynching. Nor was he as stoical as is often suggested. When his work was attacked, he would respond heatedly, sometimes publicly, and usually against his better judgment.

Moreover, Einstein's genius was less mystical than might be imagined. He was a genius—one of the finest scientific minds in history. Faced with his work, it is impossible to claim otherwise. (One of his lesser achievements, for instance, is to have theorized the process of stimulated emission, which would later become the foundation for the invention of lasers.) But he was not the archetypal, inspired, transcendental genius whose intellect is somehow apart from the world. One of the most captivating and consistent traits Einstein possessed was his ability to work—to really, truly *work* at something.

One day, when he was an assistant professor in Zurich, one of his students, Hans Tanner, came to call on him at his home. Tanner found Einstein in his study, hunched over a sprawl of papers, at work on some equations. He was writing with his right hand and holding Eduard with his left. Hans Albert, meanwhile, was giddily playing with toy bricks on the floor, trying to get his father's attention. "Wait a minute, I've nearly finished," Einstein said, handing Eduard to Tanner and turning back to his equations. Hans Albert would later recall that the sound of a baby crying never distracted Einstein. Work seems to have given him both purpose and comfort. After his first heartbreak, he wrote that "strenuous intellectual work" and the act of examining nature would between them see him through trouble and bear him through life. At other moments of extreme distress—after the death of his second wife, Elsa, or as he watched Eduard struggle with depression—he would say much the same: work was the only thing that lent significance to life.

Even within his own lifetime, Einstein was overtaken by public interpretations of him, such as the perception that he was an almost saintlike figure, with a moral superiority uncorrupted by fame. This idea was also encouraged after his death by his longtime secretary and the executor of his estate, Helen Dukas, and has firmly persisted. However, there is much to find off-putting about Einstein. As his travel diary from 1922 reveals, he harbored racist opinions of many of those he met during his tour of Asia, likewise when he toured South America in 1925. And he persistently belittled women. In his personal life, he clearly had an unpleasant strain: he was cruel to his first wife, distant as a father, and consistent in his adulteries. He also enjoyed getting his own way. He once canceled a vacation with his teenage son just because he'd dared say something Einstein wasn't happy about. He was capable of treating anything, or anyone, who constrained his sense of freedom with a petty meanness and anger.

And yet Einstein is a likable character. Some of this is due to the fun, joy, and irreverence that was part of his personality. On vacation, he would speed his boat toward other sailors on the water, only to turn away at the last minute, laughing, narrowly avoiding a collision, and this despite never having learned to swim. He called his *Autobiographical Notes*—which was the closest he ever came to writing something comprehensive about his life—his "own obituary," and then barely mentioned himself at all. Having been banned by his doctor from smoking, Einstein decided that as long as he didn't buy the tobacco himself, he was doing nothing wrong, and so he used to pilfer it from any source he could find, be it colleagues' tobacco tins or even cigarettes from the street.

Einstein probably succeeds in coming across as agreeably as he does simply because he was so friendly. Not only was he smiling and easy with strangers, but to those he liked he was doggedly loyal, affectionate, and honest. Therefore—outside his family, crucially—it is difficult to find anyone among those who knew him who was not kind about him. In Charlie Chaplin's autobiography or an interview with an aging Bertrand Russell, in the diaries of a German count or the letters of a Belgian queen, in the reminiscences of his colleagues from all of his places of work, one will consistently encounter the same feeling of happiness at having known Einstein. In the face of such affection, it becomes difficult to resist treating him as one would one's own friend: with a pleasure at being in their company, and a certain willingness to treat failures and foibles with leniency, if not forgiveness.

This book is a mosaic biography. It is composed of short chapters of varying styles that deal with a particular moment or aspect of Einstein's life—one may be an anecdote, the next a discussion of his scientific work, another a quotation from his letters. These individual pieces are intended to make up a picture, in its own way, as representative of its subject as the portrait drawn by a traditional biography. In constructing this mosaic, I don't set out to redeem Einstein, or to make a case for a defining characteristic of his personality. More fascinating, to me, are the inconsistencies inherent in a life, the inexplicable, incompatible, insane motivations that punctuate days and years.

Today, Einstein is a figurehead as much as a man, symbolic of things larger than himself: of scientific progress, the human mind, the age. He is seen as elevated by his exceptional intellect, as if he represents what we could all be capable of—an image that is added to by his outspoken righteousness, his lack of care for show or dress or awards, his indifference to what people thought of him, his resolute pursuit of truth and peace. He is, in short, a figure for good.

But examining his life shows that his genius did not overshadow his humanity, and that he was not someone formidably, dishearteningly other. When, in 1929, he published yet another attempt at a unified field theory, churches across America gave sermons on the work discussing its theological implications, and the *New York Times* sent reporters to congregations around the city. Reverend Henry Howard, pastor of the Fifth Avenue Presbyterian Church, compared Einstein's latest theory to St. Paul's preachings on the unity of nature. But the fact is that the theory wasn't a sacred text, the product of a semidivine intelligence; it was plain wrong. Albert would abandon it shortly after all the fanfare, just as he would abandon every other attempt at a unified field theory.

Einstein is a reminder that to be the best of ourselves is not to be pure beyond fault. His goodness wasn't a state of being, an aspect of genius—rather it was a pursuit. And because of that, all the more remarkable.

EINSTEIN
IN TIME
AND SPACE

1

Illustration of the Avenue de l'Opéra, Paris, 1894,
lit with Yablochkov candles.

The lights were coming on. In June 1878, a switch was thrown in Paris. The Avenue de l'Opéra—that grand road with wide sidewalks that draws the eye to the opera house—was suddenly illuminated. An unnatural, intensely bright light shot up the facades of the Haussmann architecture, leaving the upper floors in shadows. The gathered crowd gasped. The Avenue de l'Opéra was the first street in the world to be lit by electric streetlamps.

By the end of the year, these lamps, known as Yablochkov candles, had been installed on London's Thames Embankment, on posts with curling, monstrous fish at their bases. Soon their fluctuating, other-worldly glow would light every major boulevard in Paris, and thousands more would appear in London and several major cities in the United States.

Marvelous though they were, the Yablochkov candles were far too bright for indoor use, and efforts were underway to perfect an electric light bulb suitable for offices, shops, and homes. In January 1879, the British chemist Joseph Swan successfully demonstrated a working lamp at a lecture in Newcastle. The same year, in Menlo Park, New Jersey, Thomas Edison set out to perfect his version. Edison had his own glassblowing house on-site to supply him with a near constant stream of bulbs. He needed them. That year, he tested more than six thousand materials as possible filaments, which involved carbonizing almost every plant he could think of—bamboo, baywood, boxwood, cedar, hickory, flax. On October 22, 1879, a voltage was applied to a burned piece of cotton thread, coiled inside a bulb. It emitted a soft orange light and kept going for more than half a day. Edison's project had succeeded.

It was into this new, ever brightening world that Albert Einstein was born, on March 14, 1879, shortly before midday.

He was born in Ulm, an old city in Swabia—in southwest Germany—perched on the Danube. The city's motto, dating back hundreds of years, is *Ulmenses sunt mathematici*, "The people of Ulm are mathematicians." In 1805, it had been the scene of the Austrian army's defeat by Napoleon. When the Einsteins lived there, construction workers were building a steeple for the minster, where Mozart once played the organ. On completion, the church became the tallest in the world.

Pauline Einstein, eleven years younger than her husband, Hermann, was from a wealthy family. Her father, Julius Koch, ran a grain business and had managed to become "Supplier to the Royal Württembergian Court." She was educated and refined, though not considered snobbish. Well versed in German literature, she was also musical, playing the piano with both talent and enjoyment. She was said to be practical, efficient, and strong-willed, and was known for a sharp, sarcastic wit that could injure as well as cheer.

Like his wife, Hermann was descended from Jewish tradesmen and merchants. The Einsteins had made their living in rural Swabia for two centuries and with each generation had become assimilated further into German society, to the point where Hermann and Pauline were pleased to consider themselves as Swabian as they were Jewish. In fact, Einstein's parents had little interest in the Jewish religion.

Hermann was a sympathetic contrast to his wife. Easygoing, even docile, he was earthier in his tastes. He liked to walk amid good scenery; to stop at a tavern, eat sausages and radishes and drink beer. He had a walrus mustache and a square chin, and was dependably solid. At secondary school he had shown an aptitude for mathematics, and although he couldn't afford to go to university, his education had bought him entry into a higher social class. His son remembered him as wise and friendly. He was also an imperturbable optimist, even though his hopes were often wrecked by his impracticality.

In the summer of 1880, when Albert was one year old, Hermann was persuaded by his youngest brother, Jakob, to move his family to Munich and become a partner in his engineering firm, Jakob Einstein & Cie. In making the journey, the Einsteins moved from a location that was almost pastoral, where cows were still driven through the town square, to one of energetic urbanity. The capital of Bavaria was a city of three hundred thousand people. It had a university, a royal palace, and a thriving art trade.

The brothers initially dealt in water, gas, and boilermaking, but very quickly branched into electrical engineering. In 1882, they took part in the International Electrotechnical Exhibition in Munich, where they demonstrated dynamos, arc lamps, and light bulbs—and a telephone. Three years later they illuminated the Munich Oktoberfest with electric lights for the first time. So for young Albert, the electric light wasn't something abstract that suggested far-off technological revolution. It was something real, immediate, and knowable. Jakob and Hermann started to teach the boy their business. He learned about the intricacies of motors, the practicalities of electricity and light, and the physical laws that governed them.

After investing a lot of Pauline's family money, the company prospered, winning streetlamp contracts elsewhere in Germany and in northern Italy. With Jakob holding some important patents, the company employed two hundred people at its height and was able to compete with the likes of Siemens and AEG. But in 1893, when Einstein was a teenager, its fortunes changed when they lost a series of competitions to bring electric light to locations in Munich. Einstein & Cie was the only firm based in the city to compete for the contracts, but it was also the only Jewish firm, and that might have been enough to lose them business. The company went bust and Hermann and Pauline's house was repossessed. Wrenched from their home, they chose to make a new start in Italy, where business prospects were better.

Electric light surrounded the young Einstein—it was at the forefront of modern technology and at the center of the family business. But while scientists knew how to brighten town streets and make plant-fiber filaments glow gold for hours on end, light itself was still largely a mystery. That would soon change.

2

Albert and Maja Einstein, 1885.

Einstein had one sibling. His sister, two and a half years younger than him, was born in Munich on November 18, 1881. She was named Maria, although throughout her life she always used the diminutive form Maja. When Albert was informed of the imminent arrival of a baby sister he would be able to play with, he imagined something more like a toy—not the strange, small creature he encountered. On first seeing her, he asked his parents, "Yes, but where are its wheels?" He was most disappointed.

The two quickly became firm friends, however, and they remained so for the rest of their lives. Einstein's relationship with Maja was one of the most solid and loving he would experience. Their childhood was generally a comfortable one: bourgeois, easy, and happy. But Hermann and Pauline were also advocates of self-reliance, in both thinking and action, so when Einstein was three or four years old, he was sent off alone through the busiest, horse-strewn streets in Munich. He'd been shown the way once before and was now expected to manage by himself—albeit secretly shadowed by his nervous parents, who were ready to step in should anything go wrong. As it was, there was no reason to worry. When Albert reached an intersection, he would dutifully look both ways, then cross the road, completely unafraid.

In the evenings, schoolwork was to be completed before he and Maja were allowed to play any games. Young Albert would then spend his time on puzzles and with building blocks, as well as carving wood. His favorite activity was constructing houses of cards, at which he excelled, managing to build some fourteen stories tall.

Einstein's many cousins would often come to play in the family's rambling back garden, but he seldom joined in. When he did participate, he was regarded as an authority—"the obvious arbiter in all disputes," as Maja would remember. But in general, he liked his own company, was careful and thorough, and took his time about things. He developed slowly, and had been so slow in learning to talk that his worried parents consulted a doctor. He had a particular difficulty for much of his childhood: whenever he wanted to say something, he would first whisper the words to himself. This he did for every utterance, no matter how routine, which led the family maid to call him "the dopey one." Concerned about their son, Einstein's parents tried hiring a governess, who ended up nicknaming the boy "Father Bore." He finally grew out of his whispering habit at the age of seven.

Brother and sister bickered and teased each other in the normal fashion, and sometimes worse. Albert, in particular, threw violent temper tantrums when he was young, during which, as Maja recalled, his face would turn yellow and the tip of his nose white, and he would lose all control of himself. On one occasion, after he had started home-schooling, Einstein grew so incensed with the unfortunate teacher that he picked up a chair and hit her with it. She fled, never to be seen again.

"Another time he threw a large bowling ball at his little sister's head," Maja wrote some forty years later, evidently not having quite forgiven him. She also recounted an occasion when he hit her on the head with a garden hoe. "This should suffice to show that it takes a sound skull to be the sister of an intellectual."

3

One day, when he was four or five years old, Albert lay ill in bed. His father came to see him and gave him a pocket compass to examine and play with. When he studied it, Einstein became so excited that he grew cold. The needle fascinated him because he couldn't make sense of it. He knew movement could be created by contact—that was part of everyday life—but the needle was behind glass, out of reach and enclosed. Nothing was touching it, and yet it moved as if in the grip of someone's fingers.

By that age he had grown accustomed to such phenomena as the wind and rain, or the fact that the moon hung in the sky and did not fall down. They were accounted for, recognizable: they had been before his eyes since infancy. But the invariance of the compass needle, which pointed north no matter how he manipulated the case, was a wonder.

Watching the needle dance back to its position, Einstein came to understand that this was something beyond his understanding of the world. He knew nothing about Earth's magnetic fields, but it seemed to him that the needle must be influenced by some mysterious power. As he said when recalling the incident more than sixty years later, he realized that "something deeply hidden had to be behind things." And he wanted to try to understand it.

"Young as I was, the remembrance of this occurrence never left me."

4

Hermann Einstein was proud that Jewish rituals were not practiced in his home, viewing them as outdated, the remnants of "ancient superstition." In his family, just one uncle attended synagogue, and he only did so because, as he used to say, "You never know."

Therefore, when Albert turned six, his parents were happy to send him to the Petersschule, the local Catholic primary school. In his class of seventy, he was the only Jew. He received the usual Catholic education, learning sections of the catechisms, stories from the Old and New Testaments, and the sacraments. He liked these lessons, and indeed excelled at them, even to the point of helping his classmates with their work.

Einstein received no discrimination from his teachers due to his heritage. But he was, however, bullied by his fellow students, who frequently insulted and physically attacked him as he walked home from school.

To send their son to a Catholic school was one thing, to have him solely under the influence of Catholicism was another, so Albert's parents hired a distant relative to teach him the values of Judaism to act as a counterweight. Einstein, however, took things much further. In 1888, when he was nine, he suddenly developed a fervent Jewish faith. Of his own accord, he strictly adhered to dogma, obeying the strictures of the Sabbath and kosher dietary laws. He even composed his own hymns, which he sang on his way home from school. Meanwhile, his family carried on with their secular lives.

This change coincided with Albert's move to his secondary school, the Luitpold-Gymnasium, near the center of Munich. As well as paying attention to mathematics and science alongside the more traditional Latin and Greek, his new school provided a teacher to give religious instruction to its Jewish students.

Einstein later recalled finding a sort of Edenic bliss in the garden that surrounded the family house at this time. He was happy there, able to give himself to contemplation, his faith spurred by air filled with the scent of newly sprung petals, of buds and sap. He had also become conscious of what he called "the nothingness of the hopes and strivings which chase most men restlessly through life."

He referred to this phase of his life as a "religious paradise," but it ended as suddenly as it had arrived. When he was twelve, he lost all interest in religion. At that age he should have been preparing for his bar mitzvah, to make a formal commitment to Judaism, and perhaps this in itself played a part in his loss of faith. However, Einstein was later careful to attribute it to the influence of what might be called scientific thinking.

The Einsteins did keep one Jewish custom, albeit in a modified manner. It was common for Jewish families to host a poor religious student for the Sabbath meal. The Einsteins hosted a medical student on a Thursday. Max Talmud was twenty-one when he began to visit the Einsteins and Albert was ten, but the two soon became friends. After seeing his interest in the subjects, Talmud would bring Einstein science and mathematics books, and each week, Einstein would eagerly show him the problems he'd been working on. Although initially Talmud would help him, it didn't take long for Einstein to outpace him.

The effect on Einstein was profound: "Through the reading of popular scientific books, I soon reached the conviction that much in the stories of the Bible could not be true," he would recall. "The consequence was a positively fanatic [orgy of] freethinking coupled with the impression that youth is intentionally being deceived by the state through lies; it was a crushing impression."

And it was an impression he never shook off. He would forever be averse to religious orthodoxy and ritual, and hostile toward every kind of authority and dogma. An immediate upshot of this new attitude was that, at the end of three years, at the crucial moment, he refused to go ahead with the bar mitzvah.

5

Religion was not the only thing that Einstein developed an aversion to. German troops would occasionally pass through Munich, banging drums in rhythm and playing songs on fifes, stirring up a merry excitement as they went. The windows would rattle as they marched in lockstep, and children would run into the street to march with them, playing soldiers. When Einstein saw such a display once, his reaction was to burst into tears. "When I grow up," he explained to his parents, "I don't want to be one of those poor people."

This military spirit extended to education as well. At the Luitpold Gymnasium, as in most German schools of the time, the style of teaching focused on memory, discipline, and systematization. Questioning was discouraged—things were to be learned and regurgitated. Teachers were very much the center of authority and knowledge, with the student only a receptacle for that knowledge, a disciple to that authority. Einstein achieved good grades, but he was far from a good student. He was openly contemptuous of the school system, the Gymnasium, and his specific teachers, whom in later life he referred to as "lieutenants."

On one occasion, one of his teachers went so far as to declare that Einstein was unwelcome in class. He replied that he hadn't done anything wrong. "Yes, that is true," the teacher said, "but you sit there in the back row and smile, and your mere presence here undermines the respect of the class for me." The same teacher went on to say that he wished Einstein would leave the school altogether.

At the age of fifteen, Albert found himself effectively alone in Munich, forced to stay with distant relatives. After the collapse of his father's firm, the rest of the family had moved to Italy, leaving him behind to complete his education. He became so miserable that he persuaded the family doctor (an older brother of Max Talmud) to draw up a certificate stating that he was suffering from "neurological exhaustion" and claiming that he must suspend his schooling. He then went to his math teacher and asked for it to be confirmed in writing that he had mastered the subject and was an outstanding mathematician. Just before the Christmas holidays in 1894, he packed his things, bought a train ticket, and appeared, without warning, at his parents' house in Milan. Hermann and Pauline were shocked, but despite their exasperated protests, he was adamant that he would not return to Munich.

He promised that he would study independently to prepare himself for the entrance exam to the Zurich Polytechnic—the institution he had decided he would attend for his higher education. Despite their worries and doubts, in the end his parents did everything they could to help with Einstein's plan. When it was noticed that the polytechnic required applicants to be at least eighteen, Hermann and Pauline persuaded a family friend to intervene on their son's behalf and ask for an exception to be made. Their friend evidently took the task seriously, recommending the then sixteen-year-old Albert in the most excessively praiseworthy language he could think of. The director of the polytechnic, Albin Herzog, replied:

According to my experience, it is not advisable to withdraw a student from the institution in which he had begun his studies even if he is a so-called "child prodigy" . . . If you, or the relatives of the young man in question, do not share my opinion, I shall permit—under exceptional dispensation of the age stipulation—that he undergo an entrance examination in our institution.

The exam began on October 8, 1895, and lasted several days. He did not pass. While he did well in the section specific to his chosen field of study, which covered math and physics, he did poorly in everything else—the general section included the history of literature, politics, and the natural sciences. Einstein was neither confident nor idiotic enough to think that the experience went well. The glaring gaps in his knowledge must have been brought home to him, perhaps as he struggled his way through a question on zoology. "My failure," Einstein later recalled, "seemed completely justified."

Even so, thanks to his more than impressive performance in the technical section, he was encouraged by the polytechnic rather than rejected outright. The head physics professor, Heinrich Weber, invited Albert to attend his lectures, something that was against regulations. Meanwhile, Herzog recommended that Einstein should complete his final year of preliminary education at a nearby secondary school and apply again the following year. If he obtained his diploma there, Einstein would be admitted, despite still being six months below the polytechnic's age requirement.

So it was that on October 26, Einstein enrolled at the cantonal school in Aarau, a pretty town twenty-five miles from Zurich. The school had a good reputation as a forward-thinking institution. Alongside the traditional curriculum, it emphasized the teaching of modern languages and science, even having a magnificently equipped laboratory on-site. It also encouraged an accepting, positive teaching style.

Rote learning and memorization were avoided and the students were treated as individuals. In particular, it encouraged using visual images and thought experiments as a means by which to grasp concepts.

As Einstein put it, the teachers had "a simple seriousness." They were not figures of authority, but distinct people, to be conversed with, engaged with. "This school has left unforgettable impressions on me," he wrote. "Comparison with six years' schooling at a German authoritarian gymnasium made me clearly realize how superior an education towards free action and personal responsibility is to one relying on outward authority and ambition. True democracy is no empty illusion."

While in Aarau, Einstein boarded with one of the teachers at the school. Jost Winteler, his wife, Rosa, and their seven children became something close to family, and it was not long before he referred to Jost and Rosa as "Papa" and "Mama." Most evenings he would sit with them over dinner, debating and laughing.

An impressive man with a full, pointed beard, thick hair, and small spectacles, Winteler was a philologist, journalist, poet, and ornithologist in charge of Latin and Greek at the school. Generous with his time, open with his ideas, and relaxed in his teaching, he was liberal-minded, with a slightly aggressive integrity: he supported freedom of expression and held a deep-seated contempt for any form of nationalism. Einstein soon adopted many of Winteler's beliefs, noticeably his commitment to internationalism.

Einstein's newly formalized sense of politics, as well as his disdain for German militarism, made him want to renounce his nationality, and he asked his father to help him in that process. His decision was almost certainly also influenced by a rather more practical concern: if he had turned seventeen while still a German citizen, he would have been conscripted into the army.

The letter declaring Einstein officially stateless arrived six weeks before his seventeenth birthday.

6

Marie was the youngest Winteler daughter. When Einstein came to stay with the family, she was living at home, waiting to start her first job, having recently left teacher-training college. She was soon to turn eighteen; Einstein was sixteen. Full of joy and self-doubt, she was also strikingly pretty, with dark, wavy hair. They both loved music and Einstein would often play his violin for the family in the evenings, with Marie at the piano. After a few months, at the end of 1895, they fell in love.

At first, their mutual devotion was all-consuming. Einstein used to stay awake at night staring at the stars, telling himself that the constellation of Orion sparkled more beautifully than it had ever done. In January 1896, Marie moved out, to begin teaching in a nearby village. Even though she returned home often, they sent each other many love letters, lamenting their time apart: "it is beautiful to endure suffering when you console," he once wrote.

He would send her Mozart songs. He would also send her sausages in an attempt to help her gain weight, something he referred to as "the doughnut project." He would try to make her jealous, and try to make her laugh. "Guess what?" he wrote in one letter. "Today I played music with Miss Baumann . . . which you'd *have* to be envious of if you knew the girl. She can put her delicate soul into the instrument with the greatest ease, because she actually doesn't have one. Am I not nasty again and compulsively snide?"

Their parents were more than happy with the relationship. Pauline Einstein, in particular, was keen to make her happiness known. When Einstein returned to Italy in April 1896 for a spring vacation with his family, she would often make an effort to read his letters to Marie. To one of Albert's responses, Pauline attached the following note: "Without having read this letter, I send you cordial greetings!"

But it did not last. Einstein enrolled at the Zurich Polytechnic that October and settled into bohemian studenthood quickly and easily. It seems that this move almost immediately affected his attitude toward Marie, although at first he still sent her his laundry. Marie was not insensible to this. In a letter from November 1896, she wrote, with a mixture of devotion and annoyance:

> *Beloved sweetheart!*
>
> *Your little basket arrived today and in vain did I strain my eyes looking for a little note, even though the mere sight of your dear handwriting in the address was enough to make me happy . . . Last Sunday I was crossing the woods in pouring rain to take your little basket to the post office, did it arrive soon?*

Albert had already suggested that they should refrain from writing to each other anymore. "My love," was her reply, "I do not quite understand a passage in your letter. You write that you do not want to correspond with me any longer, but why not, sweetheart?" She sent him a teapot as a present. This he received with remarkable ill grace, sending a reply telling her she shouldn't have bothered. "My dear sweetheart,"

Marie answered him, "the 'matter' of my sending you the stupid little teapot does not have to please you at all as long as you are going to brew some good tea in it . . . Now, be satisfied and stop making that angry face which looked at me from all the sides and corners of the writing paper."

Albert stopped writing and she began to question their relationship, although in a characteristically backhanded way, directing much of her anger at herself. She asked whether she was not enough—she felt herself Einstein's intellectual inferior and openly thought that he was remaining with her because of some regretful obligation. For his part, Einstein seems not to have wished to cause Marie pain. He felt guilty and was still more than half in love with her, so he sought to soothe her—and extricate himself—rather than own up to his true feelings.

Eventually, in May 1897, Einstein decided to end the relationship. He sent a note in which he implored Marie not to blame herself, before going on: "I beseech you at least not to despise me because of what I, after overcoming the worst struggles, still wrested from the miserable weakling's nature. I have done nothing that deserves crushing hatred . . . only disdain."

Because he was due to visit the family shortly afterward, he was also obliged to write to Rosa Winteler, or "dear mummy," as he often addressed her:

> I cannot come to visit you at Whitsuntide. It would be more than unworthy of me to buy a few days of bliss at the cost of new pain, of which I have already caused much too much to the dear child through my fault . . . Strenuous intellectual work and looking at God's Nature are the reconciling, fortifying, yet relentlessly strict angels that shall lead me through all of life's troubles . . . And yet, what a peculiar way this is to weather the storms of life—in many a lucid moment I appear to myself as an ostrich who buries his

head in the desert sand so as not to perceive the danger. One creates a small little world for oneself, and as lamentably insignificant as it may be in comparison with the perpetually changing size of real existence, one feels miraculously great and important, just like a mole in his self-dug hole.

7

Throughout Einstein's career, he often relied on thought experiments in order to work through and describe his ideas. His writing is filled with images that explain his thinking, such as trains, embankments, and lightning strikes; a floating windowless container; blind beetles crawling across branches; an ultrasensitive device that releases one electron at a time.

Einstein was always a visual learner. In an analysis of his own thought process, he said, "The words or the language, as they are written or spoken, do not seem to play any role in my mechanism of thought." The cantonal school in Aarau specifically fostered this type of thinking, and it was during his time there, when he was sixteen, that he came up with a scenario that excited and troubled him in equal measure.

Einstein imagined a single light beam, flying through space, plunging into the darkness. He imagined someone running alongside this beam, at exactly the same speed. He realized that, in the eyes of that observer, the light would appear frozen. It would simply sit there, its peaks and troughs unmoving.

He was aware that there were problems with this, however. First, it violated a principle of science, accepted since the seventeenth century, that basically stated that the laws of physics remained the same whether an object was moving fast or slow, or staying still. According to this, the light beam should not seem to move when observed at one speed, but to stand still when observed at another.

The second difficulty was that an unmoving light wave is a light wave that effectively exists independently of time—after all, how is one to distinguish one moment from another when all is motionless? "One would arrive at a time-independent wavefield," Einstein later wrote to a friend. "But nothing like that really seems to exist!" He intuitively felt that there was something wrong about the situation.

This problem would continue to puzzle Einstein for many years, and ultimately it contained the seed of some of his greatest scientific thinking. As he would say, "This was the first childlike thought experiment in thinking about special relativity theory."

8

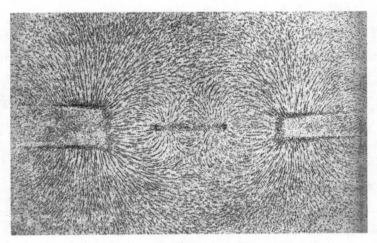

Photographic plate from the notebooks of British physicist Michael Faraday, showing the results of an experiment using iron filings to study magnetic fields generated by magnets, 1851.

On the walls of his study, Einstein liked to hang the portraits of Isaac Newton, Michael Faraday, and James Clerk Maxwell. Throughout his life they remained his scientific idols—pilots who'd set to sea before him and shown what course to take, so that he might find the shore.

What became known as the first great unification of physics was the work of Newton, who had shown, in the late 1600s, that the heavens and Earth operate under the same force of gravity—which is to say, what makes an apple fall to the ground is the same thing that keeps the moon in orbit. This was hardly obvious. Newton revealed what had seemed convincingly separate realms to be one and whole.

The second great unification of physics was the product of the other men in Einstein's picture frames. Michael Faraday was born in 1791, the son of a blacksmith. Faraday discovered that an electric current produces a magnetic field and, similarly, that a changing magnetic field produces an electric current.

Faraday was also the first person to move iron filings around with a magnet, showing how they settled into a curving pattern, stretching from pole to pole. Faraday hypothesized that a magnet sent out "lines of force," like little tendrils, to form what he called a "field." Magnetism was located not exactly in the magnet but around it, where these lines were. He speculated that it was this strange, invisible corona—the field—that influenced electric currents.

What was clear to Faraday was that electricity and magnetism were related on more than a surface level. Throughout his work, Faraday kept trying to unify what were believed to be distinct phenomena—he even speculated that light was a vibration of electromagnetic lines of force. Unfortunately, however, because of his lack of formal education, his math was very poor and so he wasn't able to supply a rigorous, theoretical justification for his beliefs.

Somebody who could do this was James Clerk Maxwell, born forty years after Faraday in Edinburgh. Like Faraday, Maxwell turned his hand to a wide range of scientific problems. It was Maxwell who was responsible for the first color photograph, for example—of a tartan ribbon. In 1859, when he was twenty-eight, Maxwell provided an explanation for the nature of Saturn's rings, something that had been bothering the scientific community for two hundred years. Even through a small telescope the rings of Saturn are visible, and it's clear that they're very thin, cosmologically speaking. The question was how the rings could be so thin and remain solid, instead of breaking up and flying into the sun or drifting off into space. Maxwell attacked the problem mathematically, showing that if the rings were solid, they would be ripped apart by gravity. Instead, he predicted that the rings were made up of vast numbers of small particles, all orbiting Saturn independently, like a giant herd. His solution was regarded as the final word on the subject. Over a hundred years later, it was confirmed by photographs taken by the *Voyager* space probes.

Maxwell is best known for his equations for electromagnetism. Maxwell had followed Faraday's work, and he set out to prove him right, aiming to provide a mathematical explanation of the link between electricity and magnetism. He succeeded, deriving four equations describing magnetic and electric fields and currents. This work was a feat in itself, but Maxwell's genius was that he realized that when he put the equations together, it suggested something new. It suggested that electric fields and magnetic fields can become caught in a feedback loop, with one producing the other. If an electric field changes in some way, it produces a changing magnetic field, which produces a changing electric field, and so on. Which is to say the two are linked in such a way that they are simply different aspects of the same thing: an electromagnetic field.

Using his equations, Maxwell was able to show that these electromagnetic fields can oscillate, creating waves. He was even able to calculate the speed of these waves. To his surprise, he found that they traveled at approximately 190,000 miles per second. This was a monumental discovery, because this was the exact speed that light had been measured to travel at. It couldn't be dismissed as a coincidence. Maxwell was compelled to conclude that light was an electromagnetic wave. And what's more, he guessed that light was part of a spectrum of such waves. Just as red light has a lower frequency than blue light, so there must be electromagnetic waves with a higher or lower frequency than light itself—Maxwell predicted the existence of radio waves more than twenty years before they were first demonstrated.

Implausible as it still sounds, Maxwell had discovered that light, electricity, and magnetism were, somehow, the same thing. This was the second great unification.

Einstein was born the year that Maxwell died and saw himself as continuing in his tradition. As he once said to a student in Berlin, "I owe more to Maxwell than to anyone." For the last thirty years of his life, Einstein would dedicate himself, fruitlessly, to discovering a "unified field theory"—a theory that unites all the disparate and fragmented aspects of physics into one harmonious whole, completing the work started by the three men on his wall.

At the time when Einstein was thinking about running alongside a light beam, Maxwell's equations were thirty years old. The landscape of physics was still in flux. It was a good time to be full of ideas.

9

Einstein in Zurich, 1898.

Zurich Polytechnic stands high in the city, on the lower slopes of the wooded Zurichberg. Beautiful in an institutional way, it is a long, imposing slab of pale sand-colored stone, situated on the wide Rämistrasse boulevard, next to the equally splendid University of Zurich. When the seventeen-year-old Einstein arrived in 1896, the small, haphazard streets that surround it would have been crammed with cafés and boardinghouses and students.

Entering the main building, Einstein would have walked into a great hallway three stories high. Regimented columns support arches and balconies, which support yet more columns, so that the hall feels almost like a monastery that has been squashed into too small a space. The doors at the far end open to reveal a spectacular view: a grassy hillside dotted with trees sloping down to the city beyond. There is the spire of Fraumünster, needle-like and teal; there the twin towers of the Grossmünster, which Wagner likened to pepper pots. In the old town there is a glimpse of a Renaissance guildhall, of banks and restaurants.

Zurich was laden with history and yet detached from it. With its cleanliness and commitment to commerce, it could be dull, but it also had a heritage of democracy and a respect for freedom. At the turn of the century, it became a haven for intellectual radicals. Rosa Luxemburg, the revolutionary who would play a key role in the foundation of the Communist Party of Germany, was already in the city when Einstein arrived. Carl Jung would move there in 1900. During the First World War, it would be in Zurich that James Joyce and Lenin would find asylum, and there, too, the Dadaists would emerge, bent on rejecting the traditions of artistic creation.

Einstein lived on one hundred francs a month—supplied by his wealthy Koch relatives rather than his father, who was in financial trouble again. He rented a room in student lodgings near the "Poly," lived off a frugal diet, and wandered the streets wearing a felt hat and smoking an outrageous yard-long pipe. He frequently studied alone and considered himself a "vagabond and a loner," but he also enjoyed the company of many friends and acquaintances in cafés or at musical evenings, where he would often play his violin.

At the time, the polytechnic was mainly a technical and teachers' college, and had fewer than a thousand students. Einstein was enrolled in Mathematical Section VIA, the "school for specialized teachers in mathematics and physics," along with ten other first-years. Among them were Louis Kollros, the son of a baker who was training to become a mathematician; Jakob Ehrat, who often sat next to Albert in class, worried a good deal, and frequently needed pep talks to get through his work; and Marcel Grossmann, a talented, diligent student, whose father owned a factory outside the city.

It didn't take long for Grossmann to announce to his parents that "Einstein will one day be a great man." Einstein visited Grossmann's family home in Thalwil on Lake Zurich, and once a week he and Marcel would drink iced coffee and smoke in the Café Metropol, where, looking out over the quayside, they would discuss philosophy and their studies. Grossmann was the type of friend very much worth having. He was liked by the teachers, attended every lecture, and wrote scrupulous notes, which he was happy to share with Albert. "His notes could have been published," Einstein later wrote. "When it came time to prepare for my exams, he would always lend me those notebooks, and they were my savior."

This is not false modesty. Unlike Grossmann, Einstein was not the ideal student. Far from it. As he admitted himself, he didn't have the talent of grasping everything easily, did not pay attention in lectures, and was less than conscientious with the work he was given.

Mostly, he struggled with math—out of 6, he scored 4 in most of his math courses, compared to 5 or 6 in physics. But then, as an opinionated young man, he had decided that higher mathematics was a waste of time. Beyond elementary math, he felt that "everything else involved subtleties that were unproductive for physicists." It was such a large subject, he believed, that he could have spent all his time, and frittered away all his energy, studying a remote speciality. Hermann Minkowski, one of his math professors whose lectures were particularly dense, said of him that "he never bothered about mathematics at all." Indeed, Minkowski remembered him as a "lazy dog."

Initially, this laziness did not extend to his physics courses. He threw himself into both theoretical and practical classes, and liked and admired his primary physics professor, Heinrich Weber—the same professor who had been so impressed with Einstein's failed entrance exam.

Einstein looked forward to Weber's lectures in a way he seldom did with any of the others.

Einstein's admiration didn't last, however—by his third year he had grown disenchanted with Weber. For one thing, Weber liked to have things just so. He once made Einstein rewrite an entire essay because the paper he'd used was not of precisely regulation size. Einstein, in turn, balked at any kind of authority, let alone an imposing one. He took to addressing Weber as "Herr Weber" instead of "Herr Professor," knowingly offending the professor's sense of propriety. By the end of Einstein's time at the polytechnic, the two were hardened enemies. "You're a very clever boy, Einstein," Weber once told him. "An extremely clever boy. But you have one great fault: you'll never let yourself be told anything."

Most of Einstein's objections stemmed from the feeling that the professor spent too much time on the history of physics, rather than exploring the subject's present and future. Weber was a representative of a not uncommon view of physics at the end of the nineteenth century: namely, that everything was basically sorted out. It was believed that Newton had elucidated the fundamental workings of the world, and since then the task of physicists had effectively been to fill in the gaps in their knowledge, to measure ever more accurately the effects of the known laws and to explain phenomena by mathematical deduction. No modern breakthroughs were discussed in Weber's classes. Anything that broke with the received wisdom of the past was ignored.

Einstein was particularly dismayed when it became clear that Weber had no intention of discussing James Clerk Maxwell's work detailing the link between electricity, magnetism, and light. Einstein was not one to hide his feelings and he let his disappointment, and disdain, be known.

Einstein and his classmates had no choice but to find out what was happening in contemporary physics for themselves. Contrary to

Weber's conservative view, evidence was mounting to suggest that the subject was inching not toward a close but a new era. In 1895, Wilhelm Röntgen had detected X-rays, mysterious things that could pass through flesh. A year later, Henri Becquerel accidentally discovered radioactivity, while J. J. Thomson discovered the electron a year after that, a tiny particle that seemed to exist within the atom. Heinrich Hertz had produced and detected radio waves in 1888, confirming Maxwell's theory of electromagnetism in the process. With the new century, Einstein and his friends were sure that marvelous new discoveries would be made. And they would be the ones to make them.

When he should have been attending lectures, Einstein studied contemporary theoretical physics "with a divine zeal" in his room. Among his copious reading, he encountered Ludwig Boltzmann, who, at a time when the existence of the atom was still in question, claimed that thinking of a gas as a whole bundle of atoms, bouncing off each other like marbles, could explain the laws of thermodynamics. Einstein also read August Föppl—who called into question the idea of "absolute motion," noting that it is only possible to define motion relative to something else—and the work of Henri Poincaré, the great French polymath, known as "the Last Universalist" of mathematics. Poincaré would be responsible for a series of insights that came very close to the concepts underpinning special relativity: he believed that, as he would write later, "absolute space, absolute time, even geometry, are not conditions to be imposed on mechanics."

Weber was not the only physics professor irritated by Einstein. Jean Pernet, who was in charge of experimental and lab work, also disliked him, and with good reason. Einstein's attendance record at Pernet's classes was so poor that he received a "reprimand from the director for lack of diligence." Einstein told one of his fellow students he thought the professor was mad, yet whenever he was in the lab, it was Einstein who was a liability. On one occasion, he threw his instruction sheet in the bin and recklessly proceeded as he liked with the experiment; on another, he caused an explosion that damaged his right hand so seriously that he had to give up the violin for a time.

For one course—in what might have seemed an act of revenge had it not been quite so justified—Pernet gave Einstein the lowest possible grade. He felt compelled to ask Einstein why he was studying physics at all, and not medicine, or law, or philology. "Because, Herr Professor, I have even less talent for those subjects. Why not at least try my luck with physics?"

In addition to their core subjects, students were required to take one extra class outside their field of study. Einstein enrolled in far more than he needed to. His eclectic choices included the geology of mountain ranges, man's prehistory, the philosophy of Immanuel Kant, the theory of scientific thought, and a study of Johann Wolfgang von Goethe's works, as well as a host of lectures on economics and politics—"Banking and Stock Exchange Dealing," "Statistics and Life Insurance," "Income Distribution and the Social Consequences of Free Competition."

Zurich was Einstein's pleasure dome. He would talk politics with his friend Friedrich Adler or spend time at the Swiss Society for Ethical Culture, where he could discuss social reform or the dangers of military might. Weekend outings often took him to the surrounding mountains and he also liked to sail on Lake Zurich. He played music at every opportunity. One summer's day, Einstein was at home with his landlady's daughter when he heard, drifting in from the house next door, the sound of one of Mozart's piano sonatas. Whipping his violin under his arm, he rushed for the door without bothering to put on a collar or tie. "You can't go like that, Herr Einstein!" his landlady's daughter called after him, but he had already gone. He rushed next door, climbed the stairs, and burst in on an old lady, who was sitting at the piano. She stared at the young man in shock. "Go on playing," was all he said. Soon enough he was accompanying her on the violin.

10

t is Saturday evening, in the months before term resumes, 1896, or maybe 1897, and tobacco smoke thickens the air of a party. Among the crowd is a short man in his early twenties. He has a dense, black, bushy beard, rough hair, and a sharp nose. He would look something like a prophet were it not for his good humor. His name is Michele Angelo Besso.

The musicians play Mozart, a little Beethoven. An appreciative audience forms a small halo around them. The violinist in particular glitters with talent. He is maybe eighteen, a student definitely, with dark hair and a neat mustache, as well as an enviable energy and confidence. After his performance, Besso can see a few people have fallen in love with the violinist. Their hostess must be pleased, he thinks; good violinists are hard to come by.

Later in the evening, somebody introduces them.

"Albert Einstein," the violinist says, smiling, without taking his pipe from his mouth.

"I'm Besso."

Einstein is gaiety itself, even if the detached amusement in his eyes never completely disappears. He's quick and funny, and conversation is easy. They soon learn that they have much in common. This Einstein is studying at the polytechnic. Besso graduated a few years ago, and now works at an electrical machinery factory in Winterthur, around ten miles from Zurich. They are both Jewish. Besso keeps up his interest in physics and is delighted to speak to someone so knowledgeable and enthused about the subject. The two men discuss whether light is a wave or a particle.

Besso and Einstein would become best friends. Like Albert, Michele had a certain disrespect for authority—he, too, had been asked to leave school when he was younger, after circulating a petition against his math teacher—as well as an open mind and a desire for foundational truths. "Besso's strength is extraordinary intelligence," Einstein wrote in 1926, "and unlimited dedication to his professional and moral obligation; his weakness is too little decisiveness."

The two friends were also both absent-minded. Einstein had an endearing, if frustrating, habit of losing his keys, but Besso could be far worse. On one occasion, he was asked to inspect some newly installed power lines just outside Milan. He set off in the evening, but missed his train. The following day he didn't remember until too late that he was meant to go at all. On the third day, while he did make it to the power station, he found that, to his horror, he had forgotten what he had been sent there to do. He had to send a postcard to his office asking for instructions to be wired to him. As Einstein put it, "Michele is a terrible schlemiel"—a bumbling, incompetent fool.

They knew and valued each other for the rest of their lives. At another party in Zurich, Einstein introduced his friend to Anna Winteler—the oldest sister of his old girlfriend, who apparently used to spy on Albert and Marie kissing. She and Michele would later marry. Only one other relationship that Einstein formed in Zurich would be more important to him.

11

Mileva Marić, 1896.

S he waited over a month to respond to his first letter.

I would have replied immediately, would have thanked you for the sacrifice involved in writing four long pages, would have also given some expression to the joy you provided me through our trip together, but you said that I should write to you some day when I happened to be bored. And I am very obedient . . . and I waited and waited for the boredom to set in; but so far my waiting has been in vain.

Mileva Marić was Serbian and had grown up in the northern province of Vojvodina, which, at the time of her birth in 1875, was part of southern Hungary. On the boundary of the Hapsburg and Ottoman empires, it was a lowland area that been populated largely by refugees and colonists. Vojvodina was a frontier, and it had given rise to the usual frontier myths: the people were determined and practical; they were wild, with something of the cowboy about them. In Mileva ran the blood of bandits, as her father used to say to her, and as she would repeat herself.

By the time she entered the polytechnic, as the only woman in Einstein's section, Marić was twenty-one, more than three years older than him. She had passed from one top school to another throughout her childhood, always excelling, especially in physics and math, and always with the support of her ambitious, loving father. Thanks to his advocacy, she had become one of the first girls in the Austro-Hungarian Empire to attend high school alongside boys.

Born with a dislocated hip, she had a limp and was commonly regarded as unattractive. "She seems to be a very good girl," one of her friends wrote in a pen portrait, "very clever and serious; small, frail, brunette, ugly." She was shy and often morose, but with a hint of grit and steel. Her intellectual intensity sometimes spilled over into personal intensity. She could be brooding and enigmatic, and she had a temper.

Throughout their first year in Zurich, she and Albert, the self-assured young German, would attend the same compulsory courses—descriptive geometry, calculus, mechanics, and so on—and somewhere along the line they grew more than passingly fond of each other. During the summer vacation in 1897, the two went hiking together.

As the new academic year began, however, Marić decided not to return to Zurich, but instead to informally attend classes at Heidelberg University. In part, this decision seems to have been arrived at because her developing feelings for Einstein had worried her: they threatened her idea of purpose. Albert took Mileva's absence as an opportunity to establish a correspondence with her.

After six months, Marić returned to Zurich. With the excuse of helping her catch up with missed work, Einstein installed himself in her company and they were soon a couple. "We understand each other's black souls so well," he wrote, "and also drinking coffee and eating sausages." They were both playful, wicked, and had the capacity for deep intimacy, while at the same time remaining detached.

Einstein loved Marić for her mind. "How proud I will be to have a little doctor for a sweetheart," he once wrote to her. Their relationship became filled with study, and linked to it. In the early days they would borrow each other's books, read textbooks together, and discuss their joint talent and passion. On one occasion, Albert, finding himself locked out of his building again, simply went to Mileva's empty flat and took her copy of a book they were reading, leaving an apologetic note that pleaded, "Don't be angry with me." When he went on a family vacation in 1899, in their third year at the polytechnic, he wrote to her imagining a perfect date they could have on his return. They would climb a mountain on the outskirts of Zurich—and then they would start to study Hermann von Helmholtz's electromagnetic theory of light.

That is not to say that theirs was a romance only of the higher realms—Einstein loved her soul, too, and her body. In one letter he signed off, "Friendly greetings, etc., especially the latter." Nor were they immune to the more domestic pleasures of companionship. Marić enjoyed looking after her dysfunctional boyfriend and he rather liked being chastised by her. In company, he would teasingly start jokes that he knew she would object to:

"Have I told you the one about the old whore . . ."

"Albert!"

And then he would burst out laughing.

Albert's friends were not impressed. They were certain he could have done better than a gloomy, ugly girl with a limp, three years his senior.

"I would never have the courage to marry a woman unless she were absolutely sound," one of them said.

"But she has such a lovely voice," Albert replied.

Mileva's friends were just as dissatisfied with the situation. Einstein may have been handsome, but he was also disheveled, with uncombed hair and holes in most of his clothes. He was absent-minded and, well, a little weird. Sometimes he seemed to go into a trance, lost in his thoughts, unaware of the inconsequential world around him. Marić fought with them about him, which hardly helped his cause.

But Einstein and Marić didn't need other people's opinions. They saw themselves as floating above them all. "We shall remain students . . . as long as we live, and shall not give a damn about the world." He began spending more and more time at her flat, so much so that his mother decided to redirect her care packages to Marić's address.

Entering into their final year, Einstein began to call her by the pet name "Doxerl" ("Dollie"), and exercised his creativity in inventing numerous others, including his "little witch," "dear kitten," "dear little angel," "little black girl," "little frog," "little right hand." Mileva took to calling him "Johonsel"—that is, "Johnnie." A letter from that time reads simply:

> *My dear Johnnie,*
>
> *Because I like you so much, and because you're so far away that I can't give you a little kiss, I'm writing this letter to ask if you like me as much as I do you? Answer me immediately.*
>
> *Thousand kisses from your*
> *Dollie*

12

After his final exams, in July 1900, Einstein made his way to Melchtal, a small village in central Switzerland. There, loaded down with physics books, he joined his family for a vacation.

Arriving at their hotel, he went to his mother's room. The conversation first turned to his grades, and he had to admit that he'd done badly. In fact, he had finished second bottom of his class, and Marić had failed to graduate at all.

"So, what will become of Dollie?" Pauline asked, innocent at least in tone.

"My wife," Einstein replied.

This provoked the reaction he had expected. His mother threw herself onto the bed, weeping, with her head buried in the pillow. When she had recovered, she started berating him—"You are ruining your future and blocking your path through life . . . That woman cannot gain entrance to a decent family"—and accused him of sleeping with Marić. Einstein furiously denied that they had been living in sin.

He was on the verge of storming out when one of his mother's good friends entered the room. Small and full of the love of life, Mrs. Bär was liked by Einstein as well as his mother. The argument, and the tense at-mosphere, evaporated instantaneously. Everyone sat down and engaged in elegant small talk—the weather was lovely, there were new guests at the spa, there were such badly behaved children around. All went down to dinner, still as if nothing had happened, and they even had some cordial late-night music. But at the very end of the evening, when they were finally alone again, Pauline and Albert resumed their fight.

Einstein's parents had never been fond of Marić, objecting to her on much the same grounds as his friends: she was ugly, older, downbeat, serious, and physically disabled. Initially, there had been reason to be-lieve that Einstein's new relationship wouldn't last long—after all, he still flirted with most girls he met. During the family summer vacation to Mettmenstetten, in 1899, when he and Mileva were most assuredly a couple, he had invited another female student from Zurich to join him. And while staying at the Hotel Paradies in the town, he had car-ried on with Anna Schmid, the seventeen-year-old sister-in-law of the proprietor, even leaving her a love poem encouraging that they kiss.

Over the next few days at Melchtal, Pauline revisited the topic—"When you're thirty, she'll be an old hag . . . She is a book like you, but you ought to have a wife"—with similar results. Once it was clear that Einstein would not be moved and that the tactic was having no effect, civility was restored.

Albert gleefully related these events to Marić in a letter, sparing her no detail. But all was well, he was saying: He had defied his mother. He had chosen *her*.

13

Einstein set about applying for jobs. He went first to the Zurich Polytechnic, to see if he might become an assistant to a professor there. It was not uncommon for graduates to be given such positions, and it seemed a logical next step. Einstein, however, had a problem. Both of his physics professors remembered him as impudent and troublingly independent. His old habit of skipping classes didn't help him, either. There was no position to be had with Weber or Pernet, and he was also turned down by one of his math professors. In fact, among his cohorts at the polytechnic, he was the only recent graduate not to be offered a job.

Over the next two years, while supporting himself as a substitute teacher or with tutoring positions, he sent out a slew of letters. "I will soon have honored every physicist from the North Sea to the southern tip of Italy with my offer," he wrote to Marić. By April 1901, he was including a postage-paid postcard in his envelopes, in the hope of receiving a response. He rarely did.

Among the scatter of his applications was a letter to Wilhelm Ostwald in Leipzig. Ostwald was one of the great scientists of the age and in 1909 would receive the Nobel Prize in Chemistry.

March 19, 1901

Esteemed Herr Professor!

Because your work on general chemistry inspired me to write the enclosed article, I am taking the liberty of sending you a copy of it. On this occasion permit me also to inquire whether you might have use for a mathematical physicist familiar with absolute measurements. If I permit myself to make such an inquiry, it is only because I am without means, and only a position of this kind would offer me the possibility of additional education.

Failing to receive a reply, he wrote again:

April 3, 1901

Esteemed Herr Professor!

A few weeks ago I took the liberty of sending you from Zurich a short paper that I published in Wiedemann's Annalen.

Because your judgment of it matters very much to me, and I am not sure whether I included my address in the letter, I am taking the liberty of sending you my address herewith.

Einstein *had* included his address in the previous letter, of course; the ruse is reasonably obvious. He was desperate and dispirited, although no more than any other job seeker. When he wrote to Ostwald, he was living with his father in Milan. Hermann Einstein was quietly, sweetly concerned for his son, so much so that in the end he decided to write to Ostwald as well.

Although Ostwald did not reply to any of these letters, nine years later he became the first person to nominate Einstein for the Nobel Prize.

14

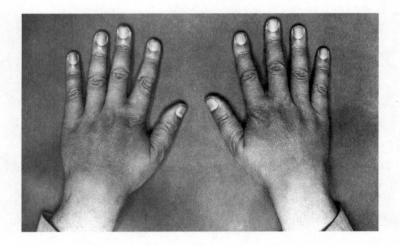

Einstein's hands, 1927.

After becoming a Swiss citizen in February 1901, Einstein, age twenty-one, presented himself for military service, as was required. His service book shows the following result of a health examination:

Body height   171.5 cm [5 ft 7½ in]
Chest circumference   87 cm [34¼ in]
Upper arm   28 cm [11 in]
Diseases or defects   *Varices* [varicose veins],
                            Pes Planus [flat feet], *Hyperidrosis ped.* [excessive foot perspiration]

He was deemed unfit to serve.

15

You absolutely must come to see me in Como, sweet little witch . . . Come to me in Como and bring along my blue dressing gown so we can wrap ourselves up in it . . . bring a happy light little heart, and a clear head. I promise you a trip more marvelous than you've ever had.

Marić hesitated briefly before excitedly accepting. They set the date for May 5, 1901, and could hardly bear the wait.

This evening I sat for two hours at the window and thought about how the law of interaction of molecular forces could be determined. I've got a very good idea . . .

Ah, writing is stupid. On Sunday I am going to kiss you in person. To a happy reunion! Greetings and hugs from your,
Albert

PS: Love!

When Mileva arrived at the train station in Como, at the very south of the Italian lake, Albert was waiting for her with "open arms and a pounding heart." They stopped there for the day, wandering the squares of the old walled town, admiring the Gothic cathedral and the medieval town hall. From there they boarded a steamship that stopped at various villages as it leisurely took them to the north end of the lake.

The next day they set out to cross the Splügen mountain pass, which rests between Italy and Switzerland. As Marić reported, it lay deep in snow, however, at some points up to twenty feet thick.

> *Therefore we rented a very small sleigh, the kind they use there, which has just enough room for two people in love with each other, and the coachman stands on a little plank in the rear and prattles all the time and calls you "signora"—could you think of anything more beautiful? . . . It was snowing so gaily all the time, and we were driving first through long galleries, then on the open road, where there was nothing but snow and more snow as far as the eye could see, so that this cold white infinity gave me the shivers and I held my sweetheart firmly in my arms under the coats and shawls with which we were covered.*

They descended on foot, having so much fun that it didn't seem like hard work at all. "In suitable places we produced avalanches so as to properly scare the world below."

A few days later Einstein wrote to Marić: "How delightful it was the last time, when I was allowed to press your dear little person to me the way nature created it, let me kiss you passionately for that, my dear, good soul!"

By the time the vacation ended, Mileva was pregnant.

16

It was only a few months before Mileva was due to retake her exams at the polytechnic. Alone in Zurich, she was bound to be scared by the implications of her situation. Einstein, she thought, could barely look after himself, let alone support her.

Albert's letter addressing her concerns began with an effusion of passion and delight—albeit regarding a physics paper he'd just read about cathode rays: "Under the influence of this beautiful piece of work I am filled with such happiness and such joy." This dealt with, he assured Marić that he would not leave her, as she had feared, but instead would make everything right. He would look for work, he told her, any work—even if it meant getting a job in the insurance business—and as soon as he was employed, they would marry.

Mileva wanted the baby to be a girl and took to calling her Lieserl, a diminutive form of Elizabeth. As Einstein wanted a boy, in mock rivalry he occasionally called the baby Hanserl. For the most part, however, Einstein didn't pay the pregnancy much attention. In his letters, he would talk of how, after the birth, he and Marić could go back to being undisturbed by the world and revel in study. "How are our little son and your doctoral thesis?" he once wrote to her.

It was understood that the two of them could not be seen together in public once Mileva's pregnancy began to show, but Einstein appears to have gone out of his way not to see her. When she invited him to visit her in Zurich, he chose instead to go on vacation with his mother and sister. In July 1901, Marić failed her exams a second time. A few months later, she went to stay at a village close to Schaffhausen, Switzerland, where Einstein was working as a private tutor, but he seems to have gone to see her remarkably little, sometimes canceling visits with the excuse that he didn't have enough money to make the short journey.

Lieserl was born around the end of January 1902, in Mileva's home city in Serbia, Novi Sad. It was a difficult birth, and Marić was seriously unwell afterward. Einstein rallied.

"It has really turned out to be a Lieserl, as you wished!" he wrote. "Is she healthy and does she cry properly? What kind of little eyes does she have? Which of us does she resemble more? . . . Is she hungry? . . . I love her so much and don't even know her yet!"

Even so, he did not visit his newborn child.

By this time Albert knew that there were problems to be overcome, not least that, before long, he was likely to be starting a new job in the city of Bern. Marcel Grossmann had persuaded his father to intervene in Albert's jobless suffering. The senior Grossmann knew the director of the Swiss Patent Office and was aware of a possible opening there. He put in a good word for Einstein. When the post of "Technical Expert III Class" was advertised, the specifications had been tailored specifically to suit Einstein. He was, the director wrote to inform him, the prime candidate.

In June 1902, the Swiss Council formally elected him to the position, with a salary of 3,500 francs, but he was unlikely to survive the probationary period if he showed up with an illegitimate child in tow. He therefore kept his daughter secret from all his friends, old and new, and when Mileva moved back to Switzerland, she didn't bring her daughter with her. In August 1903, when Lieserl was nineteen months old, she came down with scarlet fever. Mileva rushed back to her sickbed. Almost nothing concrete is known about the fate of Lieserl. She may have died, she might have been given away, but she was not to play a part in her parents' lives again. With astonishing thoroughness, Einstein and Marić wiped their daughter from history. His mother, father, and sister were completely ignorant of her, as subsequently would be his two sons. Letters from the period were destroyed. And Einstein never spoke of Lieserl again.

17

Conrad Habicht, Maurice Solovine, and Einstein, 1903.

While waiting to be offered his job at the patent office, Albert moved to Bern, taking a flat on the second floor of a thin gray building on a wide, sloping street: Gerechtigkeitsgasse. Needing to make a little money in the meantime, he placed a small advertisement in the local newspaper for private lessons in mathematics and physics, given by "Albert Einstein, holder of the fed poly. teacher's diploma."

One day during Easter vacation in 1902, there was a ring of Einstein's doorbell. "*Herein!*" he thundered, before eventually opening the door to peer at his caller. In the gloom of the dark hallway stood a man of twenty-six, not exactly handsome, but well-dressed and self-possessed, with short hair and a beard in the French style. He introduced himself as Maurice Solovine. He had been flicking through a newspaper, he explained, and, having come across Herr Einstein's notice, had promptly set off to the address it mentioned. Perhaps Einstein could teach him theoretical physics?

So began a lifelong friendship. They talked for two hours at their first session, about philosophy as much as physics, and when Solovine made to leave, Einstein accompanied him onto the street so that they could talk for half an hour more, before agreeing to meet the next day. With their second session, the physics lessons were entirely forgotten. By the third day Einstein told him, "As a matter of fact, you don't have to be tutored in physics; our discussion of problems that stem from it is much more interesting. Just come to see me and I will be glad to see you."

The two decided to read and discuss the great thinkers, and after a few weeks they were joined in their endeavors by another recent friend of Einstein's, Conrad Habicht, a banker's son. He had moved to Bern to finish his studies and was hoping to become a teacher of mathematics. Books with titles such as *What Are Numbers?* and *On the Nature of Things in Themselves* passed between the three of them. Henri Poincaré's *Science and Hypothesis* enthralled the little group for weeks. They also read plays and Charles Dickens, as well as the works of the philosophers David Hume, Baruch Spinoza, and Ernst Mach. Their sessions usually took place in one of their rooms or at the Café Bollwerk. Either way, their meals were models of frugality. "The menu ordinarily consisted of one bologna sausage, a piece of Gruyère cheese, a fruit, a small container of honey and one or two cups of tea. But our joy was boundless." Occasionally, Einstein would entertain his friends by playing the violin.

Solovine christened their little club the Olympia Academy, as a bit of a joke. Einstein, despite being the youngest, was elected president, earning him the title "Albert Ritter von Steissbein" (roughly, "Sir

Albert, Knight of Backside"). A certificate was made up, featuring a drawing of a bust of Einstein beneath a string of sausages—the dedication read:

> Expert in the noble arts, versed in all literary forms—leading the age towards learning, a man perfectly and clearly erudite, imbued with exquisite, subtle and elegant knowledge, steeped in the revolutionary science of the cosmos, bursting with knowledge of natural things, a man with the greatest peace of mind and marvelous family virtue, never shrinking from civic duties, the most powerful guide to those fabulous, receptive molecules, infallible high priest of the church of the poor in spirit.

They had fun. One day, when rambling through the arcades of the city, the three of them passed a delicatessen. On display, in among all manner of delicious-looking rare foods, was some caviar. Solovine began singing its praises.

"Is it as good as all that?" asked Einstein.

When Einstein ate a food odd or new to him, he grew completely ecstatic and would go about describing it in the most enthusiastic, fantastic words—much to the amusement of his friends. Solovine and Habicht agreed that they should save up some money, buy some caviar, and serve it to Einstein on his birthday, eager to see what he would say about it. When March 14 arrived, rather than the usual bologna sausage, Solovine filled their dishes with a mound of caviar. By the time they came to the table, Einstein was fervently discussing Galileo's principle of inertia, so wrapped up in the problem that he consumed mouthful after mouthful of caviar without mentioning it at all, continuing instead to talk about inertia. Habicht and Solovine were hardly listening—they kept glancing at each other in amazement.

Albert finished his food.

"Say," exclaimed Solovine, "do you know what you've been eating?"

"Oh, for goodness' sake," he replied, finally realizing what he had shoveled down, "it was that famous caviar."

There was a stunned silence.

"It doesn't matter," Einstein went on. "There's no point in serving the most delicious dishes to peasants; they can't appreciate them."

But his friends were determined to have him enjoy caviar. A few days later they bought some more and this time served it to him while chanting, to a melody from Beethoven's Eighth Symphony, "Now we are eating caviar . . . Now we are eating caviar . . ."

Solovine only missed a meeting once. It was his turn to host, but having spotted some surprisingly cheap tickets to a Czech quartet that evening, he decided to attend the concert instead. In recompense, he left his fellow academicians four hard-boiled eggs, knowing that they were especially fond of them, as well as a note in Latin: "*Amicis carissimis ova dura et salutem*" ("Hard eggs and a greeting to very dear friends"). He had asked his landlady to tell Einstein and Habicht that he had been called away because of some urgent matter.

They were not to be fooled. They ate the eggs and then, knowing that Solovine hated tobacco, Einstein broke out his pipe and Habicht his thick cigars, and they smoked as if possessed. The cigar butts and smoldering pipe ash they left in a saucer. For good measure, they piled Solovine's furniture and possessions on his bed until they almost reached the ceiling. "*Amico carissimo fumum spissum et salutem,*" they wrote on a piece of paper they pinned to the wall. "Thick smoke and a greeting to a very dear friend."

On summer evenings after their meetings, they would sometimes climb the Gurten, a mountain south of the city. They would walk all night, warm under the stars, discussing astronomy. "We would reach the summit at dawn, and marvel at the sun as it came slowly toward the horizon and finally appeared in all its splendor to bathe the Alps in a mystic rose." They would wait, lit in the soft orange of morning, for the summit restaurant to open, then drink dark coffee before starting their descent.

When Albert and Mileva married in Bern, in January 1903, Solovine and Habicht were the only two guests.

18

Einstein at the Bern Patent Office, ca.1905.

A DAY AT THE OFFICE, 1904

Husband and wife wish each other a good day. Their son, Hans Albert, now a few months old, is crying as Einstein leaves. He heads down the narrow staircase of their building, out into the noise of the morning. Some nice weather today, almost as if summer's returned. He's wearing a light-colored, checkered three-piece suit and a sharp silk tie. His hair is under control, after a fashion, his mustache trimmed. He looks as respectable and as kempt as is really possible for him.

He sets off down the cobbled street. Straight ahead, framed by uniform apartment buildings, is the Zytglogge, the famous clock tower over the old city gate, with its Renaissance astrolabe and minute hand decorated with the face of the sun. It's a little before eight. A few minutes later and Einstein reaches Genfergasse, near the railway station. There sits the Postal and Telegraph Administration Building, a hard, neoclassical affair, with the acceptable, self-important sterility of most new office blocks. The patent office is on the top floor.

He nods and smiles at his colleagues, says what a lovely day it is. Josef Sauter is already there, as is Michele Besso, one of his closest friends from his Zurich days, which already seem far away. Besso started at the office a few months ago, after Einstein encouraged him to apply. Maneuvering down the long room, Albert soon settles on his stool, in front of a sturdy wooden desk.

Einstein likes his work—it's diverse and stimulating. Today there is an application for a typewriter, another for a camera. He almost immediately rejects yet another—this for an electrical-engineering device—as it hasn't been drafted properly. It's the job of the patent officer to decide whether each invention is in fact new, or if it infringes on another patent. The officer also has to discern, crucially, whether the invention would actually work. Einstein's boss, Friedrich Haller, a gruff, clever, respected man, has schooled him in the proper way to deal with his forms: "When you pick up an application, think that everything the inventor says is wrong." He tells him to stay vigilant in his criticism, to refute every premise he can. The advice chimes with Einstein's way of thinking, as useful for his scientific work as it is for his day job.

He eats lunch at his desk, working out some puzzle or staring out the window. Sauter—who also studied at the polytechnic, a little before Einstein—slides over to ask if he's going to next week's Society meeting. Through his older colleague, Einstein has been introduced to the scientific circles in Bern and he often goes as his guest to meetings of the Natural Science Society. He's even been the speaker once, when he discussed electromagnetic waves, just before a talk on veterinary medicine. This month, apparently, there's going to be something about the rhinoceros. Einstein assures Sauter that he's looking forward to it.

Shortly after his break, Einstein has finished his official work for the day. He often rattles through his applications in two or three hours. This allows him to reach into his drawer and pull out his physics notes, which he surreptitiously scatters across the desktop. Rocking on his stool, he pursues his own work. When Haller passes by, he quickly crams the mess of papers back into his desk drawer while trying to appear busy. Haller politely pretends not to notice—after all, Herr Einstein is good at his job.

Not far away, the mechanical figure of Father Time appears from the Zytglogge as the clock tower rings out the hour, signaling the end of the working day. Einstein waits for Besso to gather up his things and the two talk physics as they head out together—or, more accurately, Albert talks and Michele listens, occasionally putting in a question. Einstein promises with his goodbye to have Besso and his wife over for dinner sometime soon: they had an excellent evening last time, and Anna and Mileva get on so well. Soon the street and time have passed airily under him and he is home.

Inside the flat, Hans Albert is crying again—Mileva is settling the baby. She asks about his day. His eyes absently pursuing a thought, Albert tells her what he's been discussing with Michele, before cuddling his son, who really is a handsome fellow. He talks for a little while with his wife, but he's still mulling over small-scale thermodynamics. He really should get some work done before they settle down. If he can get a few more papers published, he might start to make a bit of a name for himself.

19

Einstein's time at the patent office is often regarded as a golden incubatory period for his scientific thinking. Beyond the confines of academia, he was under no pressure to keep producing papers—he had no ladder to climb, and so his ideas could grow and develop independently and slowly. Moreover, his office work nurtured his innate critical instinct and sharpened his ability to extrapolate complex systems from visual prompts and basic premises.

But the years Einstein spent as a technical expert did not just benefit the theoretical physicist in him; they also satisfied the engineer. He had grown up around new technologies and machinery—he had worked in his father's factory and his uncle Jakob had invented and patented several electrical devices. Over his lifetime, Einstein would file numerous patents of his own.

One morning in the mid-1920s, Einstein came across an article about an entire family who had died in their beds as a result of poisonous gases leaking from their refrigerator. At that time, mechanical fridges were just beginning to enter the home (the Einsteins still used an icebox in their flat), but no nontoxic refrigerant had yet been invented. Moved by the story, Einstein wondered if he could create a safer alternative.

To help him, he recruited Leo Szilard, a brilliant young physicist at Berlin University, who went on to play a pivotal role in the development of atomic energy and become instrumental in setting up the Manhattan Project. The two men patented multiple refrigerator designs, including one that required only an external heat source to work. As part of the cooling system, Einstein designed an ingenious pump that used electro-magnetism to work. It acted like a piston, moving a liquid metal back and forth, despite having no mechanical moving parts. Unfortunately, the fridge was not very efficient and sounded a little like a wailing banshee. Despite some interest, the project never came to fruition.

Among his other inventions was a hearing aid, which he designed with the German inventor Rudolf Goldschmidt. The "electromagnetic sound apparatus" converted an acoustic signal to an electrical one, which would then feed into a membrane attached to the skull, so that the bone would conduct it to the ear. Goldschmidt and Einstein were granted a patent in 1934, but by then the rise of the Third Reich had undone their partnership and Einstein was no longer in Europe.

A year later, he was busy with various other projects, in collaboration with his friend Gustav Peter Bucky. The two invented a waterproof, breathable fabric, submitting, then withdrawing, a design for an overcoat made of threads woven so tightly that the intervening spaces between them would be the same size as waterdrops. At the same time, they were busy creating a camera that automatically adjusted to light levels. Light would hit a photoelectric cell in the camera attached to a variety of screens of different transparencies. The light intensity would determine which screen would slip down in front of the lens. As usual with Einstein's inventions, it never really attracted commercial interest. Kodak came out with their own camera with automatic exposure two years later.

20

As far as physicists in 1905 were concerned, light was a wave. There was no question about it. It was a fact, experimentally verified and at the base of over a century of theory. Whether light was emitted from a star or a glowworm, it spread out evenly through space, most definitely traveling as an electromagnetic wave.

Albert was remarkably busy in 1905. He was still working six days a week at the patent office, he had a one-year-old son to help look after, and that year he wrote twenty-one reviews for an academic journal. He also moved house in May. And yet he managed to produce five scientific papers in six months, three of which would eventually transform physics.

The first paper of his miracle year was completed in March and titled "On a Heuristic Point of View Concerning the Production and Transformation of Light." It was the most revolutionary thing he ever wrote. In it, he suggested that light should be considered as a stream of particles.

He was knowingly addressing one of the questions at the heart of modern physics: Was the world fundamentally continuous or discontinuous? Using the imagery of the philosopher and mathematician Bertrand Russell, was the world a bucket of molasses or a pail of sand? Down at the deepest level of reality, were things smooth, an unbroken, continuous whole, or were they granular, made up of particles? By a growing, though certainly disputed, consensus, matter was seen as being made up of atoms. Light, however, remained steadfastly indivisible. This disparity was troubling and bothersome, and so scientists had been directing their attention to the boundary at which the two conflicting viewpoints intersected. It was hoped that by studying the interaction of light and matter, the perplexing workings of nature might reveal themselves.

A lot of work had gone into examining what was called "blackbody radiation," a prime and promising example of just this type of interaction. As everyone knows, hot things glow red, hotter things yellow, and really hot things white. When glass is heated up, it doesn't shine forth with an encouraging green or pale chestnut; it glows red, then yellow, then white. In order to study this radiation, scientists such as the German physicist Gustav Kirchhoff had built ovens that could be heated and maintained at precise temperatures so that measurements of the different wavelengths of light emanating from their interior could be accurately recorded.

It was discovered that the wavelengths were affected only by the temperature of the oven. No matter what material was used or what shape was constructed, the same color would be achieved by the same temperature. If iron is heated to, say, 700 degrees, it radiates exactly the same spectrum of light as any other solid element at 700 degrees. The findings were indisputable, but there was a problem. No one could come up with a mathematical explanation for this behavior. No one was able to explain what it really meant. Here, light was in contact with atoms in the walls of the oven. Continuity and discontinuity stood cheek by jowl, and something, it seemed, was wrong.

The situation wasn't resolved until 1900, when Max Planck introduced the idea of the "quantum," and by doing so sent physics down the path to revolution and modernity. Planck was an unlikely radical. Descended from an old German family of lawyers, scholars, and theologians, the forty-two-year-old physics professor was about as different in attitude from Einstein as it was possible to be. Shy, formal, and correct, he was a conservative in outlook who loved and respected tradition, and a proud patriot. He was an advert for the value of diligence as well as brilliance, and he was always immaculately turned out—in his youth he wore a fine pince-nez.

Planck's conservatism extended into his science. For years he had been a virulent opponent of atoms and anything else that harmed the idea of continuous matter, but when confronting the blackbody problem, in order to make any headway, Planck was forced to embrace controversy. He succeeded—albeit after some fortuitous guesswork—in coming up with a formula that agreed with the experimental results. Strangely, however, it involved a mathematical constant—the truly Lilliputian number 6.626×10^{-34} Js, now given the symbol h—which at first Planck could not account for. He did not know what process in reality it related to.

To make sense of it, he had to assume something rather crazy. The walls of anything radiating light and heat, he argued, contained within them vibrating molecules, or "harmonic oscillators." It was the vibrations of these oscillators that produced the light appearing from the oven. For this to work, however, the oscillators could only emit and absorb energy in packets or bundles—what Planck called "quanta." Effectively, the oscillators could only emit and absorb discontinuous particles of energy. These bundles contained only specific, fixed amounts of energy; there was no sliding scale. You could have either $1 or $2, but no change in between. The energies available to the quanta were determined by Planck's constant.

Planck was sure his explanation of absorption and emission did not apply to the nature of light itself, regarding his theory as a mathematical convenience. For the purposes of blackbody radiation, energy may come in equal, finite packages—but, as everybody knew, light was a wave.

When Einstein read Planck's paper in 1901, it felt to him as if the ground "had been pulled out from under one," as he wrote later. However, after four years, he realized that it hadn't actually gone far enough. For one thing, it perpetuated the divide between continuous light and discontinuous atoms. In his March paper, Einstein argued, with intuitive but flawless reasoning, that it wasn't only the exchanges of energy between light and matter that operated in this particular, bundle-like way. The light waves themselves were made up of particles—what Einstein called "light quanta." Today, they are called photons.

Einstein went on to show the effectiveness of considering light in this way. He demonstrated that the existence of light quanta could explain something called the "photoelectric effect." It had been discovered that it was possible to knock electrons out of a metal surface with a beam of light. The energy of these ejected electrons turned out to depend entirely on the frequency (which is to say the color) of the light used. No matter how bright the light, the electrons only gained in energy if the frequency was increased; no matter how dim the light, they only decreased in energy as a result of a decrease of frequency. This was surprising, as a very bright light has more energy than a dim one.

The wave theory couldn't account for this, but Einstein was able to do so with ease. The energy of a photon is the product of the frequency of the light multiplied by Planck's constant: $E = h\nu$. Assuming, as Einstein did, that a photon "transfers its entire energy to a single electron," this means that changing the frequency directly changes how much energy a photon can impart to an electron. Making the light brighter, but doing nothing to the frequency, would produce more photons and do nothing to their energy. As a result, more electrons would be emitted from the metal, but their energies would remain the same.

Einstein expected a significant reaction to his March paper. He had, after all, just overturned one of the basic tenets of modern physics. Yet when the paper was published in the *Annalen der Physik*, it was met not with fanfare and cheers, or even with disavowal and outrage, but with silence.

21

Hardly pausing for rest, Einstein began work on a new topic, plunging once more into the world of the small. His second paper of 1905 was completed on April 30. "A New Determination of Molecular Dimensions" involved a dense mathematical examination of viscosity and the behavior of dissolved sugar molecules in water, which produced a way to determine molecular size. (A sugar molecule, he estimated, was about a millionth of a millimeter in diameter.)

Through his calculations, he was also able to estimate a value for the constant known as Avogadro's number. Hydrogen has an atomic mass of 1; helium has an atomic mass of 4; lithium has an atomic mass of 7. If I had 1 gram of hydrogen, 4 grams of helium, and 7 grams of lithium, they would all contain exactly the same number of atoms. The same with 16 grams of oxygen or 122 grams of antimony. That number of atoms is Avogadro's number. The value Einstein came up with was 210,000,000,000,000,000,000,000—or 2.1×10^{23}. It's hard to state just how big this number is. And the current, more accurate, value of Avogadro's constant is more than three times as large as Einstein's result. In the tiniest flyspeck of water reside billions upon billions of atoms.

It was this April paper that Albert chose to submit to the University of Zurich as his dissertation, in what was his third attempt to gain a doctorate. If his paper on light quanta was the most revolutionary paper of his miracle year, then this was the least. But for the purposes of a doctoral thesis, this worked in his favor—it proved he was a scientist who respected classical physics and knew how to operate within it. Not a single eyebrow would have been raised by his findings, except perhaps in admiration. That summer, after a brief correction to some of his data, his thesis was accepted, and Albert finally became Dr. Einstein.

No more than nine days after completing the second paper, Einstein produced yet another investigation of the processes of the hidden world. In this paper he set out to explain something called "Brownian motion," named after the Scottish botanist Robert Brown, who in 1828 had first investigated it. Brown had been studying pollen grains suspended in water. With his eye to the microscope, he noticed a strange phenomenon—particles within the pollen grains seemed to be dancing around at random, permanently quivering. At first, Brown took this to be a characteristic of the male sex cells and the result of life, but he decided to test his theory. He examined pollen grains that had been dead for more than a century and found the same thing. He moved on to inanimate matter, studying smoke particles, ground glass splinters, and minute chips of granite—for some reason he even tested filings from the Great Sphinx of Giza—and with all of them he saw that the particles still wandered, forever jostling as if of their own accord. Over the years, a number of explanations for this behavior had been put forward, but none turned out to be plausible.

Thanks to kinetic theory (which had been on the rise since the 1870s, and which used the random motion of molecules to explain the behavior of fluids), Einstein knew that within a liquid the molecules were not distributed evenly. Instead, they flew around at different speeds, briefly bunching in one area before dispersing and bunching in another. Sensibly enough, Einstein reasoned that this motion would affect a particle sitting within a liquid.

Even the most finely ground piece of limestone from the Great Sphinx is gargantuan in comparison to a water molecule—it's around ten thousand times as large. One water molecule would have no more chance of moving it than a pea would of shifting the Empire State Building. But immersed in water, the limestone particle was sustaining millions of collisions per second. As Einstein showed, the random distribution of these collisions would produce the jostling effect that could be observed with a microscope. At one moment, one side of the particle may be bombarded more than the other; the next moment, the side facing the heaviest volley may have switched. This would produce a random walk, with the particle moving, seeming to drunkenly stagger from place to place.

Using some of the statistics he had developed for his dissertation, Albert was able to calculate that, through all of its zigging and zagging, a standard particle suspended in water would be displaced by 0.006 millimeter in one minute. His prediction, he pointed out, could be tested. A lot hung on whether or not an experiment verified his result.

At the time, atoms and molecules were far from being regarded as real. Many physicists and chemists believed in them and they had proved to have a theoretical use. But it was still wondered whether they actually existed, or whether they were much like what Planck thought of his quanta—a convenient fiction. Einstein's predicted value for displacement was very specific, and his method of obtaining it had direct links to the science of atoms. If Einstein's result was proved right, atoms and molecules existed; if it was wrong, they didn't.

There was no shortage of responses to this paper—it garnered attention and correspondence from theorists and experimentalists alike. An attempt to verify Einstein's prediction was undertaken within months, and it was confirmed four years later, with extreme precision, by the French physicist Jean Perrin. The atomic skeptics ceded their position. Atoms were now, to all intents and purposes, conclusively real. Perrin was later awarded the Nobel Prize for his work confirming Einstein's theory.

At the end of this initial outpouring of work, Einstein took a moment to write to his friend Conrad Habicht, who had moved from Bern to Schaffhausen toward the end of 1903. They hadn't written to each other in a little while, and Einstein joked that he was almost committing sacrilege by breaking their solemn silence with the "inconsequential babble" of his letter.

"So, what are you up to, you frozen whale, you smoked, dried, canned piece of soul, or whatever else I would like to hurl at your head?" he wrote. "Why have you still not sent me your dissertation? Don't you know that I am one of the 1½ fellows who would read it with interest and pleasure, you wretched man? I promise you four papers in return." The first, he said, "deals with radiation and the energy properties of light and is very revolutionary, as you will see if you send me your work *first*." The second paper, he explained, was "a determination of the true sizes of atoms," and the third accounted for the random movements of molecules in liquid.

The last paper wasn't finished, but he was sure it would be of particular interest. "The fourth paper is only a rough draft at this point, and is an electrodynamics of moving bodies which employs a modification of the theory of space and time."

Habicht could not know it, but his friend was about to tear down another veil and force the world to be seen anew.

22

Imagine that you have shut yourself belowdecks on a ship. The main cabin is in a curious state: the air is thick with butterflies; on a table there stands a large bowl of water containing some fish; a bottle suspended from the ceiling is emptying drop by drop into a vessel beneath it. The ship is still in harbor, at rest, and so you see the fish swimming this way and that, the butterflies floating with equal speed in any direction, the drops from the bottle splashing into the vessel. Now imagine that the ship has set sail, traveling at a constant velocity on remarkably calm seas. You wouldn't be able to tell the difference. Nothing would change.

This thought experiment was described by Galileo in 1632. He used it to defend the Copernican view of the solar system from its detractors, who argued that if Earth were spinning around the sun, people would surely feel it. Galileo's ship exemplifies what is known as the "special case" of relativity. The general principle of relativity states that the laws of physics remain the same no matter what your motion, which is simple to understand, though difficult to accept. Special relativity, on the other hand, only relates to reference frames at a constant velocity, which is to say that it only relates to things either at rest or moving with a uniform speed and direction. Acceleration is not taken into account in special relativity.

Although the "special case" sounds contrived due to its strict limitations, it is much easier to grasp as a concept because we experience the effects of it in everyday life. On the Shinkansen bullet train, for example, commuters are not flung about the carriage at two hundred miles per hour, careering down the aisle, but stay comfortably in their seats. They could hold an archery tournament, bake a Breton cake, or broadcast radio waves, and the laws of physics would operate just as if they were standing on the ground.

Intrinsic to relativity is the idea that one reference frame is not privileged over another. If someone stood at a station and saw the train speed past them, then logically enough it would look to them as if the train and all its passengers were moving in a certain direction. However, to someone on the train, it would look as if *the station* were speeding past them in the opposite direction. Both of these interpretations are in fact valid. There is no way to discern which is "correct." As the laws of physics operate in exactly the same way for both the person on the train and the person at the station, no experiment can determine which of them is *truly* at rest or in motion. It all depends on your reference frame.

By 1905, the principle of relativity was a long-accepted part of physics. Einstein did not invent the idea. What separated his theory of special relativity from Galileo's ship was his consideration of light.

Sound waves wouldn't exist without something to oscillate through, such as the air or a piece of wood. Water waves wouldn't exist without water. For physicists of the early twentieth century, a wave was, by definition, a disturbance propagating through some medium. While Newton had envisioned light to be composed of particles, by the late nineteenth century it was considered instead to be a wave. James Clerk

Maxwell had brilliantly shown that light was part of a whole spectrum of electromagnetic waves, the combination of electric and magnetic fields. Moreover, light consistently acted like a wave in experiments—it diffracted and reflected, and it had a measurable frequency. It was therefore assumed that light must be propagating through some medium, like all other waves. This unknown substance was named the "ether."

For the ether to accord with observable reality, it needed to be somewhat strange. For one thing, it had to pervade the entire universe, otherwise starlight wouldn't be able to reach Earth. It also had to be so thin and spectral that it had no effect on anything within it, and yet so rigid that it allowed light to travel through it at immense speed. Much of late nineteenth-century physics was concerned with searching for this ether. It proved most elusive.

One way in which it was thought possible to discover the ether was to detect variations in the speed of light. It was assumed that Earth, moving through the ether, would create an "ether wind" that would blow in the direction opposite to Earth's motion, just as would happen if the planet were moving through air or water. Light, it was reasoned, would have a harder time traveling against the ether wind than if it were traveling with it. In 1887, the American physicists Albert Michelson and Edward Morley conducted what became a famous experiment based on this idea, splitting a light beam in two so that one half traveled with the movement of Earth and the other half traveled transverse to it. Try as they might, they could not detect the slightest difference between the speeds of the two light beams.

Of all the many experiments conducted to find evidence of the ether, none succeeded. Something was obviously wrong. However, to scientists of the late nineteenth and early twentieth centuries, the ether remained *real*—as real as air. The thought that light could propagate without a medium, through nothing, was preposterous.

By 1905, Einstein had grown skeptical of the ether's existence. In his June paper, "On the Electrodynamics of Moving Bodies," he discarded it with hardly a backward glance. "The introduction of a 'light ether' will prove to be superfluous," he wrote, doing away with two hundred years' worth of received scientific wisdom as if it were an old coat.

Einstein's paper was based on two principles only. Everything else in the theory developed directly from these immutable truths. The first was the "principle of relativity": that the laws of physics are the same in all non-accelerating reference frames. The second principle was that the speed of light traveling in empty space is constant. Light travels at 299,792,458 meters per second, or 670,616,629 miles per hour; or, if you prefer, one light-year per year. A pretty close approximation of this speed was known by the late 1800s. Einstein dared to propose that this speed remained the same no matter what the motion of a light source. In other words, the speed of light was constant in all reference frames.

If, back on our train moving with a constant velocity, you threw a ball down your passenger car, in the direction the train was moving, you would see the ball traveling with whatever speed you had thrown it at. However, someone standing at the train station, looking in through the windows at your reckless behavior as the train passed them, would see something different. They would see the ball traveling at the speed of the train plus your throwing speed. The same would be true if you threw a piece of the Great Sphinx, or indeed if you talked—the sound waves produced would seem to the person on the station platform to travel at the speed of the train plus the speed of sound.

Before Einstein, it was assumed that light would behave just like everything else. If you fired a laser down the train or held up a lantern, it was believed that, in your reference frame, the emitted light would travel at the speed of light, but for the person at the station, the light would travel at the speed of the train plus the speed of light.

Einstein took it to be a law of nature that light—unlike everything else—traveled at the same velocity for both the person on the train and the person at the station. He was sure this was correct. He was equally sure that the principle of relativity was correct, and that to develop his theory, it needed to be built up from these two postulates.

He had reached this conclusion sometime before 1905. Unfortunately for him, however, as he admitted in his June paper, these two principles were "seemingly incompatible," and as a result he had spent "almost a year fruitlessly thinking about it." Then, one beautiful day in Bern, he went to visit his good friend Michele Besso and told him about his problem. "I'm going to give it up," he said. But the friends discussed it, and all at once the ivy was pulled from the edifice and Einstein understood. The very next day Albert visited Besso again and, without any greeting, said, "Thank you. I've completely solved my problem."

And so he had. Five weeks later, at the end of June, Einstein submitted his paper. What he'd grasped in his talk with Michele was that what was a simultaneous event in one reference frame needn't be a simultaneous event in another. If a person sees two lightning strikes happening simultaneously, what that effectively means is that they are standing at a midpoint between the strikes and the light from each reaches them at the same time. But if that person were standing in the same position on a train moving toward one of the strikes, then the light from that strike would reach them before the other. One strike would happen first. Under the principle of relativity both of these views are as valid as each other. *True* simultaneity does not exist. One might say that simultaneity is a relative concept.

And what this means, as Einstein saw, is that there is no absolute time. As he put it later, "There is no audible tick-tock everywhere in the world." Every reference frame has its own time. It was this new concept of time that allowed Einstein to reconcile his two principles, and it had some strange consequences.

Einstein pointed out that if time is relative, then so is space. Imagine two people, each with their own set of synchronized clocks—one in a train, one at the station. Imagine that the train passenger has a shiny golden rod with them. To measure the length of this fantastic rod, the passenger would simply use a measuring stick. To the person at rest, however, the rod is moving. For them to be able to measure its length, a more convoluted process is required. First they need to determine where the two ends of the rod are at a particular instant in time. Once they know the position of the front and back of the rod, they can mark those points with flags and measure the distance between them. Common sense tells us that the two lengths would be the same. They're not. The golden rod appears shorter for the person at rest than for the person moving with it. The reason for this is that the person at the station has a different notion of simultaneity than the train passenger. The passenger would argue that the person at rest located the two ends of the rod at different times, not in the same instant. This phenomenon is known as "length contraction."

The other counterintuitive outcome of special relativity is called "time dilation." Imagine now that the person on the train has got rid of their rod and has upgraded it for two mirrors. One mirror is attached to the floor, the other to the ceiling. A light beam is bouncing between the two. To the train passenger, the light bounces in a straight line, up and down. But to the person at the station the light travels in a zigzag. It looks to them as if the light has to travel diagonally upward from the floor to reach the ceiling mirror, which has moved ahead a little with the train, and it travels diagonally downward to reach the floor mirror, which has moved ahead in turn. For the person at the station, then, the light seems to have traveled a greater distance than for the person on the train. But the speed of light—as is very much a given in Einstein's relativity—travels at the same speed for both observers. It can only be concluded that for the person at rest more time has elapsed than for the person on the train.

The faster the train goes, the farther the light beam has to travel from ceiling to floor. Which is another way of saying that the faster the train goes, the slower time passes. The passengers age less, plants take longer to germinate and grow, atoms decay at a slower rate. On Earth, all these effects are barely perceptible—indeed, the effects of relativity only really become interesting at very high speeds.

Einstein's June paper was unusual not only for its new and profound oddities, but also for the more mundane fact that it contained not one citation. Many of the paper's ideas were already in the scientific air of the time. George F. FitzGerald and Hendrik Lorentz, for example, both independently developed the concept of length contraction, while Henri Poincaré had questioned the concept of absolute time. But all had done so as a patch for the problems of the ether. Einstein had come to this startling vision of the world alone—or very nearly alone. "In conclusion," he wrote, "let me note that my friend and colleague M. Besso steadfastly stood by me in my work on the problem here discussed, and that I am indebted to him for many an invaluable suggestion."

After finishing his paper, Einstein took to his bed for a fortnight, while Marić checked and rechecked his work.

23

As part of his application for Swiss citizenship, Einstein officially declared himself a teetotaler. As a rule, he didn't drink alcohol. He didn't like it. Once, when offered a glass of champagne, he chose to sniff the golden, bubbling liquid and leave it at that. "I do not need wine," he said, "because my brain is acquainted with intellectual drunkenness."

After Albert completed his paper on special relativity, however, he and Mileva celebrated. During their festivities they sent a postcard to their friend Conrad Habicht, which read in its entirety:

> *Both of us, alas, dead drunk under the table. Your poor*
> *Backside and wife*

A few months later, he sent Habicht another letter. An extraordinary upshot of his theory had crossed his mind, he wrote, one that was most unexpected. It seemed that mass and energy were in some way connected. "The consideration is amusing and seductive," he wrote, "but for all I know, God Almighty might be laughing at the whole matter and might have been leading me around by the nose."

In September, Albert scribbled down a follow-up to his June paper, just three pages long. By considering a body emitting radiation first from a reference frame at rest and then from one in uniform motion, he was able to develop equations relating speed and mass, and he quickly arrived at this theorem: "If a body releases the energy L in the form of radiation, its mass decreases by L/V^2." Which is to say that "the mass of a body is a measure of its energy content." Energy and mass are in fact the same thing, in different masks.

Updating Einstein's symbols, and keeping things in their simplest form, what the special theory of relativity therefore implied was the equation $E = mc^2$. The most famous equation in all of science simply fell out of Einstein's work, an afterthought to his already miraculous year.

24

Mileva Marić, Hans Albert Einstein,
and Albert Einstein, ca.1904–5.

I n April 1906, Einstein was promoted to Technical Expert II Class and
his salary raised to 4,500 francs. He had more than proved himself as
an adept bureaucrat and talented patent examiner. What's more, as his
boss pointed out when he recommended the promotion, Einstein was
now Herr Doktor.

The family moved again, renting the top floor of a house on the
tree-lined Aegertenstrasse, where they had their own furniture and
views of the Bernese Oberland mountains. Einstein was twenty-seven,

professional, proper. Because of their move, he and Besso no longer walked home together. Habicht and Solovine had long moved away. Those dancing days were gone, and he missed them. He and Marić only socialized on Sundays.

"I am doing fine," he wrote to a friend. "I am a respectable Federal ink pisser with a decent salary."

He had time enough for fiddling on his violin and indulging in a little physics, although both activities were performed within the narrow constraints set by his two-year-old. Not that Einstein minded the tyranny of his son that much. Hans Albert had grown into "quite an imposing, impertinent fellow" and Mileva and Albert often had to check themselves from laughing at his clowning around. They were so taken with him that at home they would communicate with him only in imitation of his baby talk, and it was not until the age of five that Hans Albert learned to speak proper German.

When Einstein was not helping with household chores—chopping firewood, carrying coal, or suchlike—he would spend a good deal of time having fun with his little boy. On weekends, he would push him around in his baroque baby carriage. He would bounce him on his knee while he worked, somehow still able to perform his calculations, and would play the violin for him, to soothe him and educate him about the importance of music.

He would also build him toys. He once made a little working cable car out of matchboxes. "That was one of the nicest toys I had," Hans Albert said many years later. "Out of just a little string and matchboxes and so on, he could make the most beautiful things."

25

A handful of physicists took the theory of relativity to heart, and chief among them—rather fortunately for Einstein—was Max Planck, one of the most respected and influential physicists of the time. As soon as Einstein's paper was published, Planck gave a lecture on the new theory at the University of Berlin. Word got around and Einstein started to gain supporters—even though, as Planck wrote to him in 1907, the proponents of relativity continued to constitute "a modest little band."

Over the next few years Einstein entered into enthusiastic correspondence with physicists all over Europe. Much of the post he received was addressed to the University of Bern, as it seemed only logical to his correspondents that the man behind relativity should be employed by some kind of academic institution. "I must tell you quite frankly that I am surprised to read that you must sit in an office for eight hours a day," wrote one young physicist planning to come to Bern to help Einstein. "History is full of bad jokes."

The first to visit Albert was Planck's assistant, Max von Laue, who traveled to Bern in the summer of 1907. Waiting for Einstein in the lobby of the Postal and Telegraph building, Laue spotted a rather un-kempt, round-faced man in his late twenties, but let him walk straight past, as he couldn't reasonably believe that this was the father of rela-tivity. When Einstein eventually returned, they made each other's ac-quaintance and set off for a walk.

In the first two hours, Albert managed to overturn most of the laws of physics and talk Laue half to death. He also gave his visitor a cigar, which was so bad that Laue "accidentally" dropped it from a bridge into the river Aare. At one point, with the two of them admiring a view of the Bernese Alps, Einstein commented, "I just don't understand how people can run around all over that lot."

26

Einstein's old math professor at the Zurich Polytechnic, Hermann Minkowski, was fiercely intelligent and dapper, with his long, wave-like mustache. After reading Einstein's paper on relativity, he recalled his former student: "Oh, that Einstein," he said to his assistant with a hint of regret, "always missing lectures—I really would not have believed him capable of it!"

Nevertheless, inspired by this new physics, Minkowski created a geometric representation of the equations of special relativity—and by so doing introduced the concept we now call "space-time." He came up with a type of graph that was four-dimensional, with time acting as the fourth dimension. Graphs like this are now called "space-time diagrams." Events on one of these diagrams are portrayed by a set of four coordinates, one for each dimension.

We are actually reasonably used to four-dimensional coordinates. If one of your friends invited you to a dinner party at their New York apartment, they would need to furnish you not only with their address (which is to say, a position in three dimensions), but also with the date and time of the party.

In Minkowski's terminology, which is still used today, an event in space-time is called a "world point," and a series of consecutive events—the movement of an electron, the flight of a butterfly, the orbit of a planet—creates a "world line." Everything has its own world line.

The space-time diagrams Minkowski created are usually drawn with time as the vertical axis and with only one dimension of space as the horizontal axis, so as to make them less confusing.

If you arrived at your friend's party, and then spent all of your time standing still in the corner of the room, wishing someone would talk to you, then your world line would move forward in time, but not in space. On a space-time diagram it would be a straight vertical line. The world line of someone running away from you at a constant speed would appear as a diagonal line, moving through time and space.

These graphs provide a simple way to visualize the relativity of time, space, and simultaneity for different observers. Minkowksi also invented a way of "transforming" the graphs so that one could easily switch between the perspectives of observers. A huge benefit of Minkowski's work—of thinking in terms of space-time—was that it enabled a formal, precise, and reasonably easy way of seeing the effects of special relativity.

In 1908, Minkowski lectured on his new concept at the University of Cologne. "Gentlemen!" he began with a decent amount of showmanship. "The views of space and time which I wish to lay before you have sprung from the soil of experimental physics, and therein lies their strength. They are radical. Henceforth space by itself, and time by itself, are doomed to fade away into mere shadows, and only a kind of union of the two will preserve independent reality."

To talk about space is also to talk about time—they are as one. Whereas before one might have thought about the fabric of space stretching through the universe, the stage on which events are played, now one had to think of that fabric as space *and* time.

Einstein was unimpressed when he heard about Minkowski's interpretation of his theory. He described it as "superfluous learnedness," dismissing it with the same casualness with which he had rejected the ether.

On another occasion, he complained that since mathematicians had started bothering themselves with his theory, "I do not understand it myself anymore."

27

After five years, Einstein was determined to try once again to return to academia. At the age of twenty-eight, he had published some of the most important papers of modern science and upended classical physics, but he still hadn't managed to get an academic job.

In part, it was his own fault. In 1907, he applied for an entry-level position at the University of Bern. As part of the application, candidates were expected to submit an unpublished paper, known as a habilitation thesis. This requirement could be waived, however, if the applicant had "other outstanding achievements." Einstein believed that his achievements were indeed outstanding. The faculty committee, however, disagreed, and he did not get the post.

Swallowing his pride, he applied again, this time with the habilitation, and was accepted. Finally, eight years after graduating, he began his life as an academic. However, the position he had secured—that of privatdozent, which involved giving a few lectures and collecting fees from the attendees—was neither important nor well paid. It was so unimportant and poorly paid, in fact, that Albert couldn't afford to give up his job at the patent office. He just had more to do.

Just as Marcel Grossmann had secured him his job at the patent office, now another friend came to the rescue. In 1908, a new position at the University of Zurich was created for an associate professor in theoretical physics. This associate would serve under Alfred Kleiner, the professor whose long campaign had brought the post into being. Einstein was the obvious candidate, but Kleiner wanted the position to go to his assistant, Friedrich Adler, whom Einstein knew from his student years. In recognition of Einstein's talent, Adler persuaded his boss that Albert was far better suited for the job.

Unfortunately, when Kleiner traveled to Bern to witness one of Einstein's privatdozent lectures, to "size up the beast," as Einstein put it, he was not pleased with what he saw. Einstein did not lecture well at all, partly because he was unprepared and partly because being investigated got on his nerves. After the lecture Kleiner let Einstein know that his teaching was simply not good enough. Einstein, with apparent calmness, replied that he considered the professorship "quite unnecessary."

This was untrue. Shortly after his assessment, he was angry to learn that Kleiner's poor opinion of his teaching skills was being passed around various university physics departments in Switzerland and Germany. He feared, not unreasonably, that this report would end any chance of finding a proper academic post. He wrote to Kleiner, rebuking the professor for spreading rumors about him. Kleiner relented, saying that, if he still wanted it, the position was his as long as he could demonstrate a modicum of teaching ability.

Einstein traveled to Zurich to present another lecture for inspection. "Contrary to my habit," as he told a friend, "I lectured well." Kleiner was satisfied and a few days later officially recommended Einstein to the faculty. And yet there was still some deliberation on the part of the faculty as to whether Einstein should be given the job. Some members considered his Jewish heritage a problem, as was recorded in the minutes of a faculty meeting:

The Israelites among scholars are ascribed (in numerous cases not entirely without cause) all kinds of unpleasant peculiarities of character, such as intrusiveness, impudence and a shopkeeper's mentality in the perception of their academic position. It should be said, however, that also among the Israelites there exist men who do not exhibit a trace of these disagreeable qualities, and that it is not proper, therefore, to disqualify a man only because he happens to be a Jew . . . Therefore, neither the committee nor the faculty as a whole considered it compatible with its dignity to adopt anti-Semitism as a matter of policy.

After a secret vote—which passed with ten in favor and one abstention—Einstein was offered the job. But he declined it, as his salary, it turned out, was to be less than he was making at the patent office. Eventually, the offer was increased and accepted. In 1909, Albert Einstein, finally, became a professor.

"So," as he put it to a colleague, "now I too am an official member of the guild of whores."

28

In early May 1909, a housewife living in Basel named Anna Meyer-Schmid read in the local newspaper a notice declaring Einstein's appointment. It stirred up half-forgotten memories from ten years before. She remembered being seventeen and a summer staying at her brother-in-law's Hotel Paradies in Mettmenstetten. She remembered meeting a handsome physics student on vacation with his mother, flirting and smiling with him. She remembered the love poem he had left her.

Anna, now married to a bureaucrat, sent Professor Einstein a postcard with her congratulations. He replied immediately with a letter that walked the line between courteous and suggestive. "I probably cherish the memory of the lovely weeks that I was allowed to spend near you in the Paradies even more than you do," he wrote. "I wish you much happiness with all my heart, and am sure that you have become as exquisite and cheerful a woman today as you were a lovely and joyful young girl in those days." He let her know that he had ended up marrying Miss Marić after all, and told her that despite his name now making it into the newspapers, he was the same as ever—except that his youth had gone. In a postscript, he asked her to look him up if she was ever in Zurich, and supplied his work address.

Whatever Einstein's motivations, Meyer-Schmid seems to have understood his response as a renewal of affection. She wrote a reply, which somehow Mileva intercepted. Furious that this woman should write to Albert—and that Albert should write to this woman—Marić sent a letter to Anna's husband, claiming that she didn't know what could have possibly induced her to write a second inappropriate letter, adding, either truthfully or wishfully, that Einstein had been indignant at Anna's advances.

It was left to Einstein to smooth over the situation. He wrote to George Meyer, apologizing for the matter. He admitted that he'd been careless and had overreacted to Anna's card in such a way that it had reawakened old feelings between them; but, he said, his intent had been pure. He begged Meyer not to hold a grudge against Anna, who he stressed had done nothing but act honorably. Einstein particularly apologized for Mileva's interference: "It was wrong of my wife—and excusable only on account of extreme jealousy—to behave—without my knowledge—the way she did."

Einstein saw himself as the victim rather than the perpetrator of the incident, and the whole affair sat badly with him. It cast a shadow over his relationship with his wife. Her anger and jealousy were shown at their worst, and her protectiveness began to feel suffocating.

Only a few months after the business of Anna Meyer-Schmid was settled, Einstein was distracted by another woman: Marie Winteler, his first love. Marie had never entirely left the periphery of Einstein's life—her sister Anna was married to Michele Besso, and her brother Paul would marry Einstein's sister, Maja, in 1910—but the two hadn't corresponded for a decade. He had once admitted to Mileva, with a slightly unsettling lack of tact, that "if I saw the girl again a few times, I would surely get crazy again. I am aware of that and fear it like fire."

He had not been exaggerating. It was probably the summer of 1909 when Einstein met Marie again, when it seems she came to Bern to visit her sister and brother-in-law. She was now a teacher, age thirty-two; Einstein was thirty. Their love very quickly rekindled. They would meet in places on the outskirts of Bern—on the Gurten, the green, lush mountain south of the city; in the Bremgarten Forest; in the small town of Zollikofen—all seemingly without exciting Mileva's attention.

But perhaps Marie was used to Einstein's fickleness and was unwilling to leave herself at his mercy, or perhaps she had more scruples than he did. Either way, Einstein somehow lost her confidence. He tried and failed to see her in Zurich, waiting beneath her window, and he sent her numerous letters that went without reply. In September he wrote:

> *I continue to live with the memory of the few hours in which miserly fortune brought you to me. Otherwise, my life is as wretched as possible regarding the personal aspect. I escape the eternal longing for you only through strenuous work and rumination. So, at least tell me what rationale you have for fleeing from me as if from a leper! My only happiness would be to see you again, or to receive a brief letter . . . I am as if dead in this life filled with obligations, without love and without happiness.*

Einstein was upset for months, even after his family moved to Zurich and Mileva became pregnant again. In March 1910, he wrote to Marie once more, clearly still infatuated. He assured her that "I think of you with heartfelt love every free minute and am as unhappy as one can be. Failed love, failed life, that's how it always reverberates to me." Their hours together, he told her, signified "the high point of life."

Marie was unmoved. After some months of silence, she wrote to Einstein, most likely to inform him that she was now engaged to a watch manufacturer. The news, and Marie's overall rejection, pitched

Einstein into a morbid despair. His reply was sent only a week and a half after the birth of his second son, Eduard, while Mileva was still recovering.

> *Reading your letter, it seemed to me as if I were watching my grave being dug . . . However, I thank you and benevolent Nature one more time for giving even me, the disinherited one, a few hours of pure joy through you, fifteen years ago and last year. Now, you are a different person; what I say applies to the one that was mine. Farewell and no longer think of me, the unhappy one, rather than thinking of me with hatred and bitterness. It may seem to you that I am a traitor, but that is not true.*

29

Einstein's students at the University of Zurich were not much impressed with their disheveled new professor. His trousers were too short, his watch chain was cheap iron, and he didn't carry any lecture notes, just a strip of paper the size of a visiting card crammed with scribbles and scrawls. Nor did his lecturing style do much to straighten the collectively curled lip. Einstein fumbled his way through his topics, apparently figuring things out as he went along.

"There must be some silly mathematical transformation that I can't find for a moment," he said during one lecture. "Can any of you see it?" Silence. "Then leave a quarter of a page. We won't waste any time . . ." And on he went with his subject. Ten minutes later, he suddenly stopped himself in the middle of another point and announced, "I've got it," before filling in the missing transformation.

He soon won over his students. They quickly realized that this strange insight into a scientific thought process was far better than any polished monologue. They were privileged to watch Einstein develop his ideas during the lecture. They were able to see his tools and working technique, rather than being presented with a marvelous finished box of truths that they had no idea how to prove.

More than this, Einstein was an unusually encouraging teacher. He urged his students to ask questions, to say when they did not understand. He would regularly pause to ask if everyone was following. Students could even interrupt. During the breaks in his lectures, he would let the class gather around and talk casually with him or ask him questions. He would take the time to answer thoroughly and kindly. Sometimes, when he was excited, he would take a student by the arm to discuss a point.

For Einstein, however, a professorship—even an assistant professorship—had turned out to be far more work than he had anticipated, and he would often complain that he was too busy. The faculty had given him a grueling workload and he particularly hated his practical lessons. He confided to one student that he didn't dare pick up a piece of apparatus "for fear it might explode." Even so, despite its haphazard appearance, Einstein made sure to put a great deal of planned effort into his teaching.

At the end of his evening lectures, he would usually ask if anyone was coming along to the Café Terrasse. Students and professor would decamp, walking through the city until they reached the restaurant, which overlooked the point at which the river Limmat runs out of Lake Zurich. There they would talk until closing time about physics and mathematics, but also about personal matters. The mood was decidedly light. From time to time, the gaggle of students would feel confident enough to poke fun at their professor, although never beyond a certain point. Pupils could get into Einstein's bad graces, in which case he would generally avoid them or shoot them a glance—"a glance," as one of his students remembered, which "lashed out violently" and inflicted "spiritual punishment."

Even after the lights were dimmed and the chairs were stacked, the evening was not always over. On one occasion, he announced that he had received that morning some work from Professor Planck in which there was a mistake, and would anyone like to look at it? Two intrepid students were duly taken back to Einstein's house to examine the paper.

"See if you can spot the error while I make some coffee," he told them.

Try as they might, they could see nothing wrong at all, and said as much. The coffee now brewed and handed out, Einstein showed them what they had missed. Appropriately astounded, the students suggested writing to Planck to point out the discrepancy. It was a good idea, Einstein admitted, but they shouldn't say that there was a *mistake*, as such; they should just gently explain the correct proof. The result was correct, he told his students, even if the proof was unsound.

"The main thing," he assured them, "is the content, not the mathematics. With mathematics one can prove anything."

30

Einstein did not stay long in Zurich. In March 1911, he and Marić moved to Prague, where he had been offered a full professorship. The city, it had to be admitted, was beautiful, in its Gothic, jumbled way, but the same could not be said of the society. He wrote to Besso that they "are not people with natural sentiments," complaining about the "peculiar mixture of class-based condescension and servility, without any kind of goodwill towards their fellow men," as well as the "ostentatious luxury side by side with creeping misery on the streets."

Even in this confining atmosphere, Albert did not remain entirely without company for long. He was soon introduced to a salon of mostly Jewish intellectuals, who met above the Unicorn pharmacy in the home of their host, Berta Fanta. There they would discuss philosophy and literature, give lectures, play music, and hold elaborate costume balls.

Alongside Einstein, who was rarely without his violin, Fanta's guests included philosophers, psychologists, writers, Zionists, and musicians. Franz Kafka was an occasional attendee. At one New Year's Eve party, the society apparently performed a play cowritten by him about the unlikely topic of the philosophy of Franz Brentano. Neither Kafka nor Einstein bothered to record anything about the other.

Someone who did take more notice of Einstein was Kafka's close friend, the writer, translator, and composer Max Brod. A thin, bespectacled man with an austere face, Brod was developing a novel based on the last years of Tycho Brahe, the great Danish astronomer who had made the most accurate observations of the stars and planets of his time. When *The Redemption of Tycho Brahe* was published, Brod's character of Johannes Kepler—Brahe's young assistant, who found God in the geometry of the world, and who would go on to discover the elliptical orbits of the planets—seemed rather familiar to some.

The novel follows Brahe's efforts to create a compromise between the Ptolemaic, geocentric view of the solar system and the then new and radical Copernican, heliocentric view, which Kepler is seen to embody. Brod's Kepler is a man impervious to life as it is lived; a man of calm, of science, of truth. When the German chemist Walther Nernst read the novel, he is reported to have said to Einstein, "You are this man Kepler."

31

First Solvay Conference, Brussels, 1911.
Einstein is second from the right.

Marie Curie had been widowed in 1906 when her husband, Pierre—with whom she had shared the Nobel Prize in Physics in 1903—had been killed in an accident involving a horse-drawn cart. She was devastated, as was Pierre's protégé, Paul Langevin, a physicist who, among other achievements, developed an ingenious way to describe the acceleration of a particle in a liquid, and who would become the inventor of the first sonar device for submarine detection.

Langevin had long been trapped in a loveless marriage. He had once turned up at his office covered in bruises, confessing that his wife and mother-in-law had attacked him with an iron chair. As a result of their mutual grief, he and Curie began to spend more time together and eventually became lovers, keeping an apartment near the Sorbonne so that they could meet in secret. When Langevin's wife discovered their affair, she arranged for someone to break into the flat and steal their love letters.

Around this time, at the end of October 1911, both Curie and Langevin were invited to the First Solvay Conference, held in Brussels. Intellectually, it was a formidable gathering. A third of the twenty invitees either were or would be Nobel laureates. Curie had been the first woman to win a Nobel Prize and she was the only woman invited to the conference. Each scientist had been given one thousand francs for the pleasure of their attendance by Ernest Solvay, a Belgian chemist and industrialist who had made a fortune from inventing a process for making sodium carbonate. As he wished to see his money go to some good use, and because he also had some strange ideas about gravity that he wanted people to pay attention to, Solvay decided to fund a meeting of Europe's finest scientific minds.

Just as the conference got underway, Langevin's mother-in-law showed the stolen love letters to a journalist.

"A Story of Love," ran *Le Journal* on November 4, when Curie and Langevin were still Solvay's guests. "The fires of radium which beam so mysteriously . . . have just started a fire in the heart of one of the scientists who studies their action so devotedly; and the wife and children of this scientist are in tears." "A Romance in the Laboratory," ran the headline on the front page of another paper the following day. The couple's attendance at the Solvay Conference was spun as them running off together.

"It has been known for quite some time that Langevin wants to get divorced," Einstein explained to his friend Heinrich Zangger. "If he loves Mme Curie and she loves him, they do not have to run off, because they have plenty of opportunities to meet in Paris." Curie, he went on, "is an unpretentious, honest person with more than her fill of responsibilities and burdens. She has a sparkling intelligence, but despite her passionate nature she is not attractive enough to represent a danger to anyone."

Soon after Curie's return to Paris she learned that she had been awarded the Nobel Prize in Chemistry for her discovery of radium and polonium. This made her the first person to win two Nobel Prizes. Despite this resounding endorsement of her work, the "Curie-Langevin affair" would not go away. One paper went so far as to publish the couple's love letters—possibly tampering with them—along with a diatribe directed at Curie, in which the editor, Gustave Téry, described Langevin as a "boor and a coward." Langevin felt it was his duty to challenge Téry to a duel.

In the end, neither party fired a shot and the affair was settled, but not without adding to the negative publicity surrounding Curie. News of the duel reached the Swedish Academy, who suggested to Curie that she should not come to the ceremonies in Stockholm and that it would be better if she delayed accepting her Nobel Prize until after Langevin's divorce case. They also told her that she wouldn't have been awarded the prize at all if they'd thought the story of her affair was authentic, as they now feared it was. Curie responded that she saw no connection whatsoever between her professional and personal life. Because the prize was for her work, she would be going to Stockholm. This naturally fueled more outrage in the press.

From Prague, Einstein wrote Curie a letter of support, affronted on her behalf at the treatment she was receiving:

> *Do not laugh at me for writing to you without having anything sensible to say. But I am so enraged by the base manner in which the public is presently daring to concern itself with you that I absolutely must give vent to this feeling . . . I am impelled to tell you how much*

I have come to admire your intellect, your drive and your honesty, and that I consider myself lucky to have made your personal acquaintance in Brussels. Anyone who does not number among these reptiles is certainly happy, now as before, that we have such personages among us as you, and Langevin too, real people with whom one feels privileged to be in contact. If the rabble continues to occupy itself with you, then simply don't read that hogwash, but rather leave it to the reptiles for whom it has been fabricated.

32

Imagine a person waking up in a small room, like the inside of an elevator, with no windows or doors. There is no way for them to tell where they are. All they know is that they're floating in the middle of the room. This means that either they're somewhere deep in space, away from any gravitational influence, or they're on Earth, in free fall— but they can't tell which. In each worrisome situation they would experience exactly the same sensation of weightlessness.

In November 1907, while at work at the patent office, Einstein had struck on this notion of weightlessness. Sitting at his desk, but decidedly not doing his job, it suddenly occurred to him that if a person were falling freely, they would not feel their own weight. He would later call this the "happiest thought" of his life. He was startled and excited by it, as he realized that this was the key to developing a theory of gravity. Thinking again about this unfortunate falling person, he reflected on another simple but important fact: when a person falls, they are accelerating. The insight Einstein took from this was that a person falling in a gravitational field and a person accelerating in the absence of gravity are equivalent to each other.

This equivalence is not immediately obvious, but Einstein came up with another thought experiment involving a person trapped in an elevator that helps to clarify things. This time, when they wake in their windowless room, they are not floating between the floor and ceiling, like an astronaut. Instead, their feet are planted on the floor. They could be on Earth, at rest within the planet's gravitational field, but it could still be the case that they are in space. If someone had attached a rope to the top of the chamber and was pulling it "upward" with constant acceleration, then the person inside would feel their feet pressed against the floor, just as they do on Earth. Not only would they stand as they would stand in a house or a bank or a post office, but they could also drop a cannonball and it would fall to the floor. Einstein realized that the effects of acceleration and of being in a gravitational field are indistinguishable, and he was ingenious enough to glean from this that gravity and acceleration are in fact deeply related.

Once Einstein had settled on his "equivalence principle," it guided his thinking about gravity and his efforts to generalize the special theory of relativity. He was aware that his theory of 1905 was incomplete. It was, after all, a special case, relevant only to things traveling at a constant speed in one direction or staying at rest. Einstein wanted, and felt assured that there should be, a broader theory, one that would include acceleration. "I decided to generalize the theory of relativity from systems moving with constant velocity to accelerated systems," he summarized later. "I expected that this generalization would also allow me to solve the problem of gravitation."

One counterintuitive result of the equivalence principle was that gravity should bend light. Think of our person in the elevator, accelerating upward. There is now a pinhole in the wall and a beam of light is shining through it. By the time the light has reached the opposite wall of the elevator, the elevator has moved upward and the light is closer to the floor than when it shot through the pinhole. If you drew a diagram of the light's path, it would curve downward. As the effects of gravity and acceleration should be the same, according to the equivalence principle, then light should also seem to bend when in a gravitational field.

In 1911, near the end of his time in Prague, Einstein came back to general relativity. Two phenomena held his attention. The first was the bending of light, which he had already conceived of but now gave full thought to. Light, of course, famously goes in a straight line, never taking a detour on the shortest path from A to B. Everyone knew that. How could it be, then, that light curved?

A potential answer was to think of the path of a light beam much like the shortest path between two points on the globe, or some other curved or warped surface. On these surfaces, the shortest path from A to B isn't straight, it's curved, and it is given a special name: a geodesic. It could be that the medium through which light travels, space itself, is curved by the presence of gravity. If so, it would only be fitting for light to travel in a curved geodesic rather than the more familiar straight line.

Einstein was also occupied with considering what happens when a disk rotates. When placed in the frame of special relativity, a spinning disk causes a problem. For someone in a reference frame not rotating with the disk, the circumference of the disk will contract, in much the same way that the length of a train will contract if it speeds past them as they stand at the station. This is an upshot of the constancy of the speed of light, as Einstein had worked out in special relativity, and not troublesome in itself. The problem was that, for this observer, the *diameter* of the spinning disk would not change, in the same way that the width of the train viewed by the person on the platform would not change even as its length did. Length contraction occurs only in the direction of motion. If the disk's circumference changes and its diameter doesn't, then the relationship between them cannot be defined by pi.

That a circle's diameter and circumference may be defined by pi is one of the tenets of the geometry we are used to. The Greek mathematician Euclid set it down in around 300 BC and it's proved pretty useful since then. Euclid's geometry describes shapes as they behave on flat surfaces. Within this framework the angles of a square are each 90 degrees, the angles of a triangle total 180 degrees, and so on. But Euclidean geometry could not describe what was happening to Einstein's rotating disk. If it couldn't describe rotation, then it couldn't define acceleration, as rotation is in fact a type of acceleration. And, due to the equivalence principle, if Euclidean geometry could not describe acceleration, then it couldn't describe gravity, either.

Bearing in mind the curvature of light and his thinking about rotation, Einstein realized that to generalize his theory of relativity—to have it describe acceleration and gravity—he would need to phrase it in the language of non-Euclidean geometry. This geometry describes shapes as they behave on surfaces that are not flat, but curved or warped. In a non-Euclidean space, the angles of a square won't each be 90 degrees. A triangle drawn on the surface of Earth, with lines that are the shortest paths between its points, will seem to bulge slightly compared to the traditional idea of a triangle. Its angles will total more than 180 degrees.

Unfortunately for Albert, the math involved was difficult and foreign to him, and he hadn't paid nearly enough attention to it when he was a student. But he did happen to know someone who was very good at it, indeed.

In July 1912, Einstein moved back to Zurich, having been offered a professorship at the university, and found himself back in the company of his old friend and classmate Marcel Grossmann. Grossmann was now head of the math department at the polytechnic, and he knew an awful lot about non-Euclidean geometry. It had been the subject of his dissertation, and he had since published seven papers on the topic.

Einstein called on his friend nearly as soon as he arrived in the city. "Grossmann," he said to him, "you've got to help me or I shall go crazy!" He explained his predicament, and Grossmann, aflame with enthusiasm, agreed to help. In particular, he directed Einstein to the work of Bernhard Riemann.

Riemann was one of the greatest mathematicians to have lived. The study of non-Euclidean geometry before Riemann had involved developing mathematical ways of describing the surface of spheres and other, slightly more complicated curved objects. Riemann developed a way of describing a surface even if its geometry changed from point to point—even if it was curved here, then suddenly flat there, and then, as suddenly again, warped in a completely odd way. More than this, Riemann also found a way of describing the geometry of four-dimensional space. The tool he used to do this is called a "tensor." And it was tensors that Einstein needed to grapple with if he was going to generalize relativity.

Tensors are complicated. They contain information about something. Think of a vector. A vector contains two pieces of information about something: its direction and its magnitude (which is how much of that thing there is; how much distance or how much speed, for example). The vector itself is a combination of these two factors. As a cannonball leaves the mouth of a cannon it will have a certain speed and a certain direction, and therefore one will be able to calculate a vector for the ball at that point. As the cannonball arcs into the rush of the oncoming enemy, its speed and direction will constantly change, and so there will be a vector to describe each point of the cannonball's path. A vector is a type of tensor—a pleasantly simple type. But there are tensors that contain far more than two pieces of information. And the more pieces of information tensors contain, the more difficult they are to calculate. Unsurprisingly, the tensors Einstein was using to try to figure out the makeup of the universe contained an awful lot of information.

By the time he was working with Grossmann, Einstein had already had one of the most profound thoughts in the history of science. Namely, he had realized that gravity was geometry. Gravity *was* the curvature of space-time. Space-time is, as its name suggests, space and time combined—it is the fabric of space, the medium in which everything exists. Mass warps space-time; the larger the mass, the more space-time warps. The surface of a trampoline will sink more under the weight of a bowling ball, say, than that of a glass marble. Roughly speaking, so it is with the fabric of space. Massive objects curve and distort space-time around themselves, and the more massive they are, the more space-time curves—and this is the same as saying that the more massive an object, the stronger its gravitational field.

For months after returning to Zurich, Einstein tried to develop a set of equations using tensors that would describe space-time. It was exceptionally hard work. Sometimes he would try to approach things purely mathematically, making sure that the tensors made sense in their own terms, and sometimes he would approach things with physics at the front of his mind. He wanted to ensure that his equations actually related to the world and not only to abstract math, just as much as he wanted to make sure that the math was coherent.

And by the end of 1912 he had, in fact, developed a beautiful set of equations with a tensor that was basically right. Three years before he would announce his theory to the world, he had effectively reached the correct solution to the workings of the universe. But having come up with these equations, he then sat down to test them. When describing certain circumstances, such as those on Earth, they should agree with the theory of Newton. If they didn't, then no matter how beautiful they were, they were no good. And while Einstein was checking them, he made a mistake. It seemed that this solution didn't match up with Newton after all, and so he rejected it and moved on to something else.

In 1913, Einstein and Grossmann presented some of their findings in a paper. They knew it was only a preliminary sketch—a "draft," they called it. More than incomplete, however, it was, to put it bluntly, wrong. One of the theory's main failings was that it was not "generally covariant," which is to say that the theory's equations could potentially change depending on how a person was moving. Initially, Einstein had wanted the laws of his theory to be unchanging, the same for everybody, no matter if they stood still or accelerated, or the way in which they accelerated. For another thing, the theory didn't account for the odd orbit of Mercury.

Physicists had known that Mercury's orbit was troublesome since the 1840s. The "perihelion" of a planet's orbit is the point at which it is closest to the sun. Mercury's perihelion had shifted more than was accounted for by the laws of Newton. A tiny, tiny amount more— forty-three seconds of an arc every century, to be exact—but still big enough to need accounting for. Initially it was thought that an unseen planet was pulling at Mercury, in the same way that Neptune pulled at Uranus, and this planet was given the name Vulcan. But, of course, Vulcan was not to be found—it did not exist.

If Einstein's equations were correct, they should correctly predict the deviation of Mercury's perihelion. Einstein knew this and was eager to try the calculation. He even enlisted the help of his old friend Michele Besso, who visited him in the summer, to help him run the numbers. Unfortunately, the result they came up with was nowhere near the actual value. But Einstein was still confident he was right, even without any evidence to support him. Even so, something nagged at him—there was still something lacking.

"Nature shows us only the tail of the lion," he wrote to a friend. "But there is no doubt in my mind that the lion belongs with it even if he cannot reveal himself to the eye all at once because of his huge dimensions. We see him only the way a louse that sits upon him would do."

33

Elsa and Einstein in Washington, DC, 1921.

Elsa Löwenthal had known Albert all her life. Their mothers were sisters and their fathers were cousins, so she was Einstein's first and second cousin. They spoke the same dialect, knew the same places, had access to the hoard of family stories and shared childhood memories. They had attended their first artistic experience together, at the Munich opera. Like her cousin, Elsa had grown up, moved away, and married. She'd had two daughters with a textile merchant who had frittered away their money. After twelve years of marriage, they divorced in 1908, and she moved back in with her parents.

Elsa had made herself a figure in the Berlin social scene. She occasionally gave dramatic poetry recitals in the theater. She was determined, strong-willed, and unabashedly a social climber. Despite her bad eyesight, she refused to wear glasses, and at one dinner party she started to eat a flower arrangement, mistaking it for a salad.

In April 1912, Einstein visited Berlin to call on some friends and colleagues. He also paid an obligatory visit to his aunt and uncle, where his acquaintance with Elsa was reestablished.

After Albert's return to Prague, she secretly sent him a letter. He replied breathlessly. "I can't even begin to tell you how fond I have become of you during these few days . . . I am in seventh heaven when I think of our trip to Wannsee. What I wouldn't give to repeat it!"

Within a week, though, he had begun to feel that their new relationship was wrong and sent her a letter to say as much. And he wrote again two weeks later, trying to convince himself of what he had said.

I am writing so late because I have misgivings about our affair. I have the feeling that it will not be good for the two of us as well as for the others if we form a closer attachment. So, I am writing to you today for the last time and am submitting again to the inevitable, and you must do the same . . . I bear my cross without hope.

But he could not quite bring himself to break the ties to his cousin completely. As soon as Einstein moved back to Zurich in July, he sent Elsa his new work address.

In July 1913, Einstein received a very generous job offer to become, all at once, a professor at the University of Berlin, a director of a new physics institute, and the youngest-ever member of the Prussian Academy of Sciences. There were no teaching commitments, and there was a lot of money involved. Mileva, he told Elsa, had mixed feelings about a move to Berlin: "She is afraid of the relatives, probably most so of you (rightly so, I hope!). But you and I can very well be happy with each other without her having to be hurt. You cannot take away from her something she does not possess."

He and Marić had drifted apart. "I treat my wife as an employee I cannot fire," he wrote. Love and warmth and a defense against the buffets of daily life he found instead in Elsa. "It is almost disgraceful that I am sitting down again to write to you, seeing that I received your letter only today," he wrote in October 1913. "But the hours I spent so comfortably with you left me with such a longing for pleasant conversation and cosy togetherness that I cannot resist reaching for the miserable paper surrogate of reality. Add to this that the situation in my house is ghastlier than ever: icy silence."

Elsa sent him food and bought him presents, notably a hairbrush, or as Einstein referred to it, his "bristly girlfriend." He sent mock progress reports to appease her: "The hairbrush is being regularly applied, other cleansing also relatively regular. Otherwise, conduct so so la la. Toothbrush retired."

How nice it would be, he wrote to her, if they could live a life without grandeur in a small bohemian household. This had been his dream as a student, shared with Marić, but it was far from what Elsa wanted, and cleanliness was still a point of contention:

> But if I were to start taking care of my grooming, I would no longer be my own self . . . So, to hell with it. If you find me so unappetizing, then look for a friend who is more palatable to female tastes. But I will stick to my indolence, which surely has the advantage that I am left in peace by many a "fop" who would otherwise come to see me.
>
> So, a foul profanity and a hand kiss from a hygienic distance from your really filthy
> Albert

34

I have firmly decided to bite the dust with a minimum of medical assistance when my time has come, and till then to sin to my wicked heart's desire," Einstein wrote to Elsa in 1913. "Diet: smoke like a chimney, work like a horse, eat without thinking and choosing, go for a walk *only* in really pleasant company, and thus only rarely . . . sleep irregularly."

He is teasing Elsa—for worrying over him and sending him sensible advice. The list may be a joke, but it is a knowing one and serves as a pretty accurate description of how Einstein lived his life. He smoked almost constantly. He smoked cigars—the fattest, longest ones he could find—but preferred his pipe. He believed that pipe-smoking contributed to a "somewhat calm and objective judgment in all human affairs," as he wrote on receiving lifetime membership to the Montreal Pipe Smokers Club in 1950.

His pipe was a tool to set his mind in action and help him think through his work. One of Einstein's many pipes—a bashed-about, plain thing, made from briar wood—is now housed in the National Museum of American History, where it is the single most requested object in the modern physics collection. Even in those periods when he had given up smoking, he used to carry it around with him and chew on it, so much so that he wore a hole right through the bit.

Einstein never managed to quit for long, though. In response to one of Elsa's many reminders that it was bad for his health, he once bet her that he could go without smoking from Thanksgiving to the New Year. He won the bet, but he got up at daylight on New Year's morning and didn't take his pipe out of his mouth all day except to eat.

His favorite brand of tobacco was Revelation, but anything would do. Toward the end of his life, having been forbidden from smoking once again by his doctor, Einstein changed his walk to work. His usual route took him through a green, clipped meadow. However, he found that, if instead he walked along the street, he would come across an abundance of cigar and cigarette butts that he could pick up, fill his pipe with, and smoke. He did not want to offend his doctor by openly buying tobacco. Einstein continued to smoke these cigarette butts until a friend of his agreed to keep him regularly supplied with tobacco. As long as the tobacco was not technically his, Einstein felt entitled to smoke it.

35

The Einsteins' marriage had long been deteriorating, and the move to Berlin in April 1914 effectively made a ruin of it. Marić was miserable and Einstein saw his family as an inconvenient distraction from his work. He was in love with his cousin, and Mileva had started at least one of her own affairs.

After barely two months in their new apartment, she took the boys and moved in with a mutual friend, although she still hoped that things could be rectified. Shortly afterward, Einstein sent Marić a stark memorandum setting out the conditions she would have to satisfy for them to return to living together. Not preceded by any formalities, it read:

A. You will make sure:
1. that my clothes and laundry are kept in good order;
2. that I will receive my three meals regularly *in my room*;
3. that my bedroom and study are kept neat, and especially that my desk is left for *my use only*.

B. You will renounce all personal relations with me insofar as they are not completely necessary for social reasons. Specifically, you will forgo:
   1. my sitting at home with you;
   2. my going out or traveling with you.

C. You will obey the following points in your relations with me:
   1. you will not expect any intimacy from me, nor will you reproach me in any way;
   2. you will stop talking to me if I request it;
   3. you will leave my bedroom or study immediately without protest if I request it.

D. You will undertake not to belittle me in front of our children, either through words or behavior.

This list was followed by a bitter and slightly confused postscript, in which Einstein wrote:

> *Go your own way, let yourself be deceived. I really don't care.*
> *Read this slowly. It will do you good. Read it also to your family,*
> *they have nothing else to do.*

As a note to Hans Albert, he added, "Ever since coming to Berlin you have become quite lazy."

When Marić accepted these demands, Einstein wrote to her again to make sure that she was "completely clear about the situation."

> *I'm prepared to return to our apartment, because I don't want*
> *to lose the children and because I don't want them to lose me, and*

for this reason alone. *After all that has happened, a comradely relationship with you is out of the question. It should become a loyal business relationship; the personal aspects must be reduced to a tiny remnant. In return, however, I assure you of proper comportment on my part, such as I would exercise towards any woman as a stranger. My confidence in you suffices for this, but only for this. If it is impossible for you to continue living together on this basis, I shall resign myself to the necessity of a separation.*

On July 24, 1914, they met to formalize a separation agreement. He agreed to pay Mileva around half his salary and did not demand a divorce. The following Wednesday, Marić and the two boys left Berlin for Zurich. Einstein accompanied them to the morning train and, for one of the exceptionally few times in his life, wept—in fact, he cried all afternoon and evening. Life apart from his sons was difficult to imagine and he was crushed by the prospect.

"I would be a real monster," he wrote, "if I felt any other way."

36

Extremely well connected, the philosopher and mathematician Bertrand Russell made the acquaintance of almost everyone who was anyone in the late 1800s and the first half of the twentieth century, and he seldom had a good word for any of them.

He was, for example, "less impressed by Lenin than I expected to be." In his opinion, Alfred, Lord Tennyson was "rather a fraud. I thought him a humbug." Dwight D. Eisenhower was "a silly man," while George Bernard Shaw "felt vanity, and that was the whole of him." As for D. H. Lawrence, he was "perfectly intolerable. He was a fascist."

Einstein, however, whom he knew fairly well, he found to be, as he said in 1961, a "quite unbelievably lovable man. He was extremely simple, absolutely devoid of the slightest pretension. You might have met him in a railway train and not known that he was distinguished at all. He was very, very friendly, very open-minded. I thought him altogether quite delightful—I think the most satisfactory great man that I have ever known. I couldn't have wished him different in any way."

37

Europe had been engaged in what Einstein called "saber rattling" for several years before the First World War started in 1914. Like many others, he thought it would come to nothing.

His close friend Fritz Haber, the director of the Kaiser Wilhelm Institute for Physical Chemistry in Berlin, did not run off to the front with the onset of war, but saw himself as a soldier nevertheless. His face was crossed with dueling scars from his youth. After applying for an officer's commission, he wore his military uniform to work every day and was known to wear his helmet out and about.

Haber contributed two major scientific innovations to the German war effort. The first of these has changed the very makeup of the planet. Britain quickly set up a naval blockade of Germany and its allies, hoping to control not only the food supply, but also, crucially, the nitrate supply. Nitrates were an essential ingredient in making fertilizer—and an essential component in explosives. Without nitrates, the war would have been over for Germany very quickly. Haber had spotted this potential problem even before the war and had managed to create a method to synthesize ammonia, a compound that, after a simple chemical reaction, turned into a nitrate. For the first time in history, humans were able to create agricultural fertilizer—as well as, of course, bombs—without being hampered by irregular, unreliable natural processes.

Haber's second contribution to the war was chemical weapons. He pioneered the use of both chlorine gas—which, when breathed in, combines with moisture to fill the lungs with hydrochloric acid, creating the feeling that one is both burning and drowning from the inside—and mustard gas, which causes the skin and lungs to blister, as well as temporary blindness.

Throughout and after the war, he and Einstein remained very close, despite their political beliefs and actions being totally at odds. It was with the Habers that Marić and the children had stayed during their separation. When Haber was experimenting with chlorine, Albert was tutoring his friend's son in mathematics.

Einstein was a pacifist and, at the outbreak of the war, he admitted to feeling out of place: "At such a time as this," he wrote to a friend, "one realizes what a deplorable species of brutes to which one belongs. I . . . feel a mixture of pity and revulsion."

His Swiss citizenship, which he had retained, meant that he couldn't be called on to fight for Germany, even if the kaiser had wanted such a thing, but he was in no hurry to volunteer his scientific talents. Instead, as the conflict went on, he served on the Great Council of the Central Organization for a Durable Peace and took part in meetings of the German Peace Society. He was also an early member of the New Fatherland League, which advocated setting up a European federation so as to avoid future conflicts, and which was banned by the government in 1916.

Asked to contribute to a collection of essays from intellectuals defending the war's righteousness, his entry decried the state of modern times and the biologically "aggressive characteristics of the male creature," calling war the worst enemy of human development. "But why so many words when I can say it in one sentence," he concluded, "and in a sentence very appropriate for a Jew: Honor your Master Jesus Christ not only in words and songs, but rather foremost by your deeds."

In 1918, the military authorities finally ruled to restrict Einstein's freedom of movement. It was decided he had grown too dangerous. For his wartime efforts, meanwhile, Fritz Haber went on to receive the Nobel Prize for Chemistry.

38

In Berlin, Einstein continued the battle with his gravitational field equations. It was not easy work and his struggle lasted over a year, for the most part because the path he was traveling down was not the right one. After Marić and his sons left, he moved to another apartment in a new building, which had seven rooms and not nearly enough furniture to fill them. Here, he could work, alone, at whatever time pleased him, and eat and sleep when he remembered. As he wandered through the near-empty rooms, lost in thought, more and more problems with his and Marcel Grossmann's working version of general relativity became apparent. But Einstein was not prepared to give up on it, not yet.

In the summer of 1915, he traveled to Göttingen to deliver a week-long series of lectures on his progress. The University of Göttingen was known as the foremost institution for the more mathematical aspect of theoretical physics, and one of the most eminent mathematicians on its staff was David Hilbert. Einstein was aware of Hilbert's talent, and the two struck up a friendship. Elated at having found someone who could understand what he was doing, Einstein explained the sophisticated details of his theory. Hilbert was taken with the idea—so taken, as it happened, that he determined to see if he could figure out the correct field equations himself.

Three months after his visit to Göttingen, Einstein's "draft" version of his theory had suffered another two setbacks, and finally he could no longer stand in the face of its many failings. Just as he was about to consign his years of work to the shelf, he learned of Hilbert's efforts to find a correct set of equations. He returned to the task of reviewing tensors and trying out and tweaking equations. This time, he wanted to make absolutely sure that the equations he produced were covariant, that the laws of the universe were the same for everybody under whatever circumstance.

At the same time as he was busy trying to formulate the correct set of equations to encapsulate general relativity, he was also giving a series of lectures at the Prussian Academy, in which he was summarizing the state of his theory. Every Thursday in November 1915, Einstein would arrive at the grand hall of the Royal Library to address fellow members of the academy. In his first lecture he spelled out what he had done wrong so far, admitting his complete loss of confidence in his efforts and his recent change of tack.

Throughout this period, Einstein and Hilbert corresponded, each trying to outpace the other. It was after his second lecture that Einstein received a worrying letter. Hilbert said he was ready to lay out his whole theory in complete detail on the coming Tuesday and he invited Einstein to the talk. Einstein, understandably, replied that he was unable to go—although he did anxiously ask for a proof of Hilbert's paper. At the same time, Einstein had decided to see whether his new set of equations—which he knew were not quite ready— could hold up to some tests. He redid his and Besso's old calculations regarding the orbit of Mercury, and, to his utter delight, he found that he got the right answer, to a tee: a deviation of forty-three arc seconds per century.

"I was beside myself with joy and excitement for days," he wrote. Einstein's friend Abraham Pais later remarked, "This was, I believe, by far the strongest emotional experience in Einstein's scientific life, perhaps in all his life."

He presented his findings to the academy on November 18. The same day, he received the proof of Hilbert's paper, and to his dismay found that Hilbert's work was remarkably similar to his own. Two days later, Hilbert submitted his paper to a science journal under the somewhat pretentious title "The Foundations of Physics."

Einstein, by this time exhausted and suffering severe stomach pains, managed to complete his gravitational field equations in time for his final lecture at the Royal Library on November 25. These were covariant, just as Einstein had always wanted. He had his general theory of relativity. In one of its simplest variations, where much is compressed into a neat shorthand, it reads: $G_\mu v = 8\pi T_\mu v$.

Hilbert corrected his paper before it was eventually published in December, changing his equations so that they matched the ones Einstein presented in his lecture. He never claimed priority for the discovery of the equations, and before long he and Einstein were on friendly terms once more, much to the happiness of both.

The gravitational field equations simultaneously invented and accurately described a new conception of time, space, and gravity. For all their difficulty, they reduce to say something simple. They tell us that matter defines how space-time can curve (this is represented by the left-hand side of the equation), and that curved space-time defines how matter can move (the right-hand side). So it is that the universe is in a marvelous waltz, where each partner has influence over the other and shapes the dance.

The equations also act as a practical tool. They can be used to describe any number of specific interactions between objects in space. As long as the right numbers are plugged into them, they can describe the behavior of a meteorite shooting through empty space, the path of a small planet around a giant star, the propagation of gravitational waves, the rotation of galaxies.

39

Sixteen years later, at lunch with Charlie Chaplin in California, Elsa painted a simpler picture of the moment Einstein hit upon his theory, adjusting the story to suit her taste. There was no need to bother Mr. Chaplin with all the struggles Albert had gone through trying to figure the thing out, for example, and he certainly didn't need to know that they had not been living together at the time. This is how Chaplin remembered the occasion:

At dinner, she told me the story of the morning he conceived the theory of relativity.

"The Doctor came down in his dressing gown as usual for breakfast but he hardly touched a thing. I thought something was wrong, so I asked what was troubling him. 'Darling,' he said, 'I have a wonderful idea.' And after drinking his coffee, he went to the piano and started playing. Now and again he would stop, making a few notes, then repeat, 'I've got a wonderful idea, a marvelous idea!'

"I said, 'Then for goodness' sake tell me what it is, don't keep me in suspense.'

"He said, 'It's difficult, I still have to work it out.'"

She told me he continued playing the piano and making notes for about half an hour, then went upstairs to his study, telling her that he did not wish to be disturbed, and remained there for two weeks. "Each day I sent him up his meals," she said, "and in the evening he would walk a little for exercise, then return to his work again.

"Eventually," she said, "he came down from his study looking very pale. 'That's it,' he told me, wearily putting two sheets of paper on the table. And that was his theory of relativity."

40

Title page of the Illustrierte Kronen Zeitung, *October 22, 1916,
showing the assassination of Karl von Stürgkh.*

In Vienna, at around two o'clock in the afternoon on October 21, 1916, Einstein's friend Friedrich Adler entered the Hotel Meissl & Schadn, where the minister-president of Austria, Count Karl von Stürgkh, was lunching. Adler approached Stürgkh's table, shouted, "Down with absolutism! We want peace!" and shot him three times in the head. He didn't resist arrest.

Stürgkh had been the head of the Austrian government since 1912 and had established an authoritarian, military regime, dissolving parliament and governing by decree. Adler had left Zurich for Austria the same year, giving up on an academic career in order to become a Social Democrat politician alongside his father. His planned demonstration the day before the assassination, calling for parliament to be reconvened, had been banned by Stürgkh.

When Einstein read about the murder in a Berlin newspaper, he was quite unaware of Adler's motives. Nevertheless, he was sure he had to help him. For one thing, it was Adler who had stepped aside for Einstein to take his first proper academic appointment in Zurich. Albert wrote to Adler's wife, Kathia, offering his services: "He is one of the most excellent and purest men I know," he declared. "I do not believe he would act rashly; he is much too conscientious for that." Albert also wrote to Adler in prison, offering to be a witness for the defense. Einstein then looked for help from his friends in Zurich, former colleagues of Adler's, and as a result of his efforts, the members of the Zurich Physical Society issued an authoritative, positive testimony of Adler's character.

When Adler was sentenced to death in May 1917, Einstein quickly gave an interview to a newspaper, hoping to show a favorable side to his friend, in which he went so far as to suggest Adler may have been justified in what he did: "The objectivity reflected in his scientific work," he said, "also governed his actions."

By the time of the sentencing, it was known by those in certain circles that Adler's life would probably be spared because of the deep respect commanded by his father, the leader of the Social Democrats. Through appeals, Adler's death sentence was reduced to eighteen years of "fortress detention," and he was granted an amnesty after the end of the First World War.

Even toward the end of his life, Einstein applauded the Austrians for not having put his friend to death.

41

On the Russian front, ensconced in a deep trench, sits a man over forty years old with a high forehead, a large mustache, and the marks of a lieutenant on his uniform. Having volunteered for the army, Karl Schwarzschild was first put to work in Belgium, providing weather forecasts. Here, in January 1916, his job is to calculate the trajectories of artillery shells, but for the time being he's writing a letter to Albert Einstein. He has some news. A former professor at Göttingen and the director of the Astrophysical Observatory in Potsdam, Schwarzschild is a gifted mathematician, and despite the freezing cold and the terrors of war, he has somehow managed to solve Einstein's field equations.

Even Einstein had not managed to solve his equations—indeed, he initially didn't think they *could* be solved. When trying them out for himself, he had used an approximation, in a perfectly acceptable practice, and thought that this was all that could be achieved. Differential equations are not the type of math most people are used to. For some it's quite possible for there not to be an exact solution. Trying to solve them is, at the very least, hard going. But Schwarzschild succeeded, and in only a matter of months. He did what Einstein could not. "The war is kindly disposed towards me," he had written to Einstein a couple of months earlier, "allowing me, despite fierce gunfire at a decidedly terrestrial distance, to take this walk into this your land of ideas."

Einstein's field equations describe the rules of general relativity, the set of laws by which space-time abides. A solution to the field equations describes how space-time acts in a particular set of circumstances. One solution to Einstein's equations could describe what would happen if two stars orbited each other, while another what would happen if a very large planet collided with two small moons, and so on. Understandably, Schwarzschild made his task as easy as he could for himself by addressing the simplest case possible: a nonrotating, homogeneous, spherically symmetric, non–electrically charged mass—which is to say, a ball of mass, sitting by itself. His solution describes the exact shape space-time assumes around the ball, and so allows for someone to work out the trajectories of objects moving within its vicinity. This was actually rather useful, as most of the objects out in space look basically like a sphere.

As Schwarzschild wrote, however, there seemed to be, not exactly a problem, but something unusual about his solution. It was unquestionably right, but it implied that—or rather it insisted that—if the mass of this ball was compressed into a small enough space, the calculations wouldn't work anymore. If the sun, for example, was crushed in some giant cosmic vise so that all its mass fitted within a particular radius, mathematically it would collapse into a rush of infinites. What size this radius is depends on the original mass of the object. It is now called the Schwarzschild radius, and the more massive the original object, the larger its Schwarzschild radius. The sun's is less than two miles, which is very small indeed compared to the size of the sun. At the center of this dense pocket of mass, space-time would curve infinitely in on itself. Which is to say that, within this two-mile radius, nothing would be able to escape, no item, not even light—the object that the sun has now formed is utterly black. On top of that, due to immense warping of space-time within the radius, time would seem to stop completely.

All in all, it didn't seem to make much sense, but neither correspondent was particularly bothered at the time. It was odd, yes, but not important. Einstein would always believe this mathematical quirk didn't truly represent anything in the real world. But it does. From the 1960s on, such densely packed, mathematically confusing phenomena have been known as "black holes," and they have become an increasingly important part of astronomy. It is by staring into their consuming abysses that the latest generation of physicists look for the secrets of the universe. In 2019, over a hundred years after Schwarzschild had hit on their existence, the first photograph of a black hole was taken. With all their paradoxical qualities, they are real. In fact, they are very common, spread throughout the universe, small, large, at the center of galaxies and colliding with one another.

Schwarzschild had little time to dwell on the implications of his find. While at the front, he contracted pemphigus, a rare autoimmune disease. He died a few months after sending his paper to Einstein. An asteroid, an observatory, and a crater on the dark side of the moon are named after him—as well as, of course, a type of black hole.

42

The sky is dark at night. But *why* is it dark? It's filled with billions and billions of stars, so many, in fact, that one shines at every point in the sky. We're surrounded by beacons of nuclear fusion, and the light they emit travels uninterrupted for millions of years in the vacuum of space. If in every direction there are stars, the night sky should burn white. But instead we see pinpricks in the darkness.

For thousands of years, philosophers and physicists puzzled over this question. The ancient Greeks wondered about it; so did the German astronomer Johannes Kepler in the early 1600s and Edmond Halley, in England, a century later. In 1901, the British physicist Lord Kelvin published a short paper attempting a solution, but the answer eluded him. To make sense of it, a new model of the universe was required. The blackness of the night sky is tightly bound up with one particular consequence of Einstein's general theory, and with what became known as his "greatest mistake."

General relativity—as Einstein crystallized it in his field equations in 1915—was something beautiful. It reconceived the cosmos. It explained mysteries such as the irregular orbit of Mercury and insisted on new ways of understanding time, space, and mass. But there seemed to be something wrong with it.

The problem was this: the matter within the universe—that is to say, the combined mass of all the stars and planets, meteorites and comets, everything—affects the size of the universe as a whole. When studying his equations, Einstein found that when he applied them to the entire universe, they implied that the universe could not maintain a fixed size. If the universe were static, then the gravitational forces would eventually pull all matter together. Yet, this obviously wasn't happening—everything was *not* being pulled together. And so the universe couldn't be static. If even the most basic tenets of general relativity were correct, the universe had to be either expanding or shrinking. Either the matter in the universe was causing the fabric of space to fly apart, meaning that one day the stars, the great "mansions built by Nature's hand," would be so far away from one another that blackness would rule almost everywhere. Or it was causing it to collapse inward, so that eventually all would be drawn together, to be shrunk and crunched into a single point and nothingness.

Einstein found neither of these possibilities appealing. But more important, they didn't match the evidence. In the early twentieth century, astronomers had no sense that the universe was any bigger than the Milky Way. And in our local galaxy, everything seemed to be stable. The Earth went round the sun, of course, and similar planetary motion occurred everywhere else. But, in the general cosmic scheme of things, as far as anyone could tell, things stayed in their proper places, neither receding nor advancing. The consensus among physicists of the time, including Einstein, was that the cosmos was static, with neither a beginning nor an end.

Instead of trusting in his theory, no matter how strange its conclusions, Einstein bowed to conventional wisdom and accepted that he was not quite right. He was nearly forty, at the midpoint of his life, neither young enough to reimagine the universe for a second time, nor old enough to believe he knew better than everyone else. He was a lauded professor, justly respected in the scientific community and known within Germany. He was comfortable, established, and—although he might not have liked to admit it to himself—bourgeois. Wrestling with general relativity had convinced him of the power of mathematics and had reinforced his trust in intuition and mental invention, but he still appreciated observable facts. And as far as facts went, it seemed he had little choice. He needed to adjust his equations. If he didn't, then general relativity would be nonsense, untethered from reality. It helped that Einstein truly believed that the universe was homogeneous and isotropic—which is to say that it looked the same in all directions and that Earth occupied no special place within it. He was also convinced that the universe was immutable and eternal, even if his theory implied otherwise.

In 1917, he made what he called "a slight modification" to his field equations. He added the "cosmological constant"—the *kosmologische Glied*—represented by the Greek letter Λ, lambda. It was a fix that Einstein introduced into his equations to make the universe mathematically static. This sounds like cheating, and partially it was, although it wasn't quite that bad. He didn't just tack on any old number that would give him what he wanted. First, the additional term didn't affect how well the equations worked. And, importantly, it already existed within his original framework, ready to be used—he had assumed that its value was zero and so could be left out. Now, it seemed, that was no longer the case.

What the cosmological constant did was to provide an outward push to counteract the gravitational pull of matter and stabilize the universe. Antigravity, in a word. Even though Einstein decided that Λ was required, he still wasn't happy about it. In the paper in which he introduced the constant, he said, "In order to arrive at this consistent view, we admittedly had to introduce an extension of the field equations of gravitation which is not justified by our actual knowledge of gravitation." He sounds almost dejected.

This new term ruined what he thought of as the elegance of his original equations. It was "gravely detrimental to the formal beauty" of relativity. It was tacked on, yes, but more concerning to Einstein, it *felt* tacked on. It smacked of shoddy workmanship. Not only on Einstein's part, but on God's as well. Einstein believed that the universe did not tolerate inelegance or complexity where simplicity would do. The inviolable laws that produced the universe had no business being messy. As he put it in a lecture from 1933, "Nature is the realization of the simplest conceivable mathematical concepts." Even so, although Albert never really liked the cosmological constant, he became attached to the term pretty quickly, as it did still ensure that the universe was static.

Over the next decade, during which Einstein became the most famous scientist since Newton and was awarded the Nobel Prize, the constant began to be challenged. Physicists who had studied Einstein's equations tried to persuade him that an expanding universe was more than a possibility. It was a reasonable, even probable, outcome of his theory. Einstein batted them away, although the evidence suggested they might be right.

In 1912—before Einstein had finished formulating his equations— the astronomer Vesto Slipher observed distant objects in the universe and found that they appeared to be receding. Because of his equipment, he could not definitively tie his results to an expanding universe, but in the 1920s other observations confirmed Slipher's suspicions. From the Mount Wilson Observatory in the San Gabriel Mountains above Pasadena, California—then home to the world's largest telescope—came thrilling discoveries about the outer reaches of the known universe. Their data was fragmentary, but astronomers there also saw that distant star clusters appeared to be flying away from us.

By then, however, Einstein was the most illustrious scientist in the world and had spent ten years backing the cosmological constant and a static universe. It would take something indisputable to make him change his mind. That something came in 1929, with the publication of a paper by the distinguished American astronomer Edwin Hubble, the director of Mount Wilson. The paper provided conclusive evidence for an expanding universe. In 1924, Hubble had found a galaxy outside the Milky Way: Andromeda. Soon he would find two dozen more, completely changing our conception of the size and nature of the universe. We were not one collection of stars, but rather one of many of what Immanuel Kant called "island universes," separated by huge distances.

With these new extragalactic stars available to him, Hubble and his colleague Milton L. Humason set about measuring their redshift. Redshift is to light what the Doppler effect is to sound. The frequency of a wave is altered as it moves respective to whatever receives it. If an ambulance siren or the low whistle of a freight train moves toward you, the sound waves produced will compress, meaning they will have a higher frequency and as a result will sound higher-pitched. As the train whistle moves away from you, the waves will stretch out, the frequency will lower, and so it will sound lower-pitched. It is the same with light. As a light source—say a star—moves toward an observer, the waves of light compress, producing a higher-frequency color, shifting toward the blue end of the spectrum. If the star is moving away from the observer, the frequency of the light decreases and it becomes redder.

Slipher had already seen light from distant stars shifting toward red—as had the Belgian physicist and priest Georges Lemaître. But Hubble and Humason were now able to produce substantial—unignorable—evidence that the stars in all directions were traveling away, getting redder and redder. Unless Earth was at the center of all creation, this could only mean that the universe was expanding.

The news traveled to Berlin, where it reached Einstein, who almost immediately deleted the cosmological constant from his thinking. Two years later, on his second trip to America, he made a journey with Elsa to Mount Wilson. Thanks to his celebrity, much fuss was made of the visit, not least by Hubble himself. Einstein was shown around and allowed to twiddle some dials on the famous telescope. Elsa, when she was told that all of the mighty instruments could determine the scope and shape of the universe, is reported to have replied, "Well, my husband does that on the back of an old envelope."

Albert was shown the photographic plates that Humason and Hubble had worked from and it was apparent immediately that all was in order with them. The next day, at a press conference in the observatory library, Einstein formally abandoned the constant and acknowledged that the universe was most likely not static after all.

So the universe is expanding. Say you draw some dots, all evenly spaced out, on a deflated balloon. Then you inflate the balloon. As the balloon expands, the dots will spread out, but you'll notice that the farther one dot is away from another, the faster it moves away from that dot. That's to say, if a dot is far away to begin with, the distance it retreats by when the balloon inflates will be large. If it's close, the distance will be small. This, in a nutshell, is the expanding universe, with the balloon as the universe itself and the dots as matter within it. It means that the most distant stars, as they move ever farther from us, will become dimmer and dimmer, then finally disappear. It will take age upon age—well beyond the death of the solar system—but eventually all of space will be an ocean of orphaned stars, sitting alone, in darkness. It may not be a jolly thought, but Einstein was elated.

As a by-product of this discovery, the mystery of the night sky was finally explained. An expanding universe plays two important roles in making sure there is blackness as well as light. First, as we have seen, expansion means that the light emitted by a star shifts toward the red end of the spectrum. But the process does not stop with visible light. The more distant a star, the more its electromagnetic waves are stretched out. Through this process, visible light becomes infrared, and lost to our eyes.

More important is the fact that an expanding universe presupposes a *beginning*, a point from which the expansion started. We now know this as the big bang. This happened approximately 13.8 billion years ago. And that means light has only had that amount of time to travel. Now, 13.8 billion years is certainly a very long time, but because light doesn't move from A to B instantaneously, but travels at a particular speed, for some stars it's not actually long enough for their light to have covered the distance to Earth. Moreover, because the universe is expanding, the distance the starlight has to travel keeps increasing, so that it will never reach us. The night sky is dark because the universe has a beginning.

From 1931, Einstein was joyously free of the cosmological constant, never to return to it except in derision. It served no use now. Beauty was restored.

Much later, however, it turned out this wasn't exactly true. The constant would be resurrected, to serve an unexpected purpose. It has since proved increasingly difficult to get rid of. But for the rest of Einstein's life, there it rested: the great mistake.

43

Ilse Einstein, 1921.

Ilse Einstein was twenty-one. Einstein had recently hired Elsa's older daughter as the secretary for the physics institute of which he was director. Ilse was idealistic, very left-wing, politically active, and graceful. She had lost the use of one of her dark eyes in a childhood accident, but this was taken as an attractive imperfection, and added to, rather than diminished, her charisma.

She had fallen in love with Georg Nicolai, a long-standing family friend and physiologist two decades older than her. Pacifist and cultured, having lived in both Russia and France, but with an egotistical streak, Nicolai was known for his acts of derring-do. On one famous occasion, having been court-martialed while working as an army doc-tor, he succeeded in stealing a biplane from the German air force and flying to Denmark. Tied to this sense of self-importance was his sexual appetite. On one trip to Russia, he kept a notebook in which he listed the women he'd had sex with—the tally came to one hundred and six, including two mother-daughter combinations.

In May 1918, Ilse sent Nicolai a long, rambling letter. She added the instruction "Please destroy this letter immediately after reading it!"

Yesterday, the question was suddenly raised about whether A. wished to marry Mama or me. This question, initially posed half in jest, became within a few minutes a serious matter . . . Albert himself is refusing to take any decision, he is prepared to marry either me or Mama. I know that A. loves me very much, perhaps more than any other man ever will, he also told me so himself yesterday. On the one hand, he might even prefer me as his wife, since I am young and he could have children with me, which naturally does not apply at all in Mama's case; but he is far too decent and loves Mama too much ever to mention it. You know how I stand with A. I love him very much; I have the greatest respect for him as a person. If ever there was true friendship and camaraderie between two beings of different types, those are quite certainly my feelings for A. I have never wished nor felt the least desire to be close to him physically. This is otherwise in his case—recently at least . . . The third person still to be mentioned in this odd and certainly also highly comical affair would be Mother. For the present—because she does not yet firmly believe that I am really serious—she has allowed me to choose completely freely. If she saw that I could really be happy only with A., she would surely step aside out of love for me. But it would certainly be bitterly hard for her . . . It will seem peculiar to you that I, a silly little thing of a 20-year-old, should have to decide on such a serious matter; I can hardly believe it myself and feel very unhappy doing so as well. Help me!

Einstein and Marić formally divorced in February 1919. He married Elsa in June.

44

Since arriving in Berlin, Einstein had steadily become more accepting, more pleased even, with the idea of belonging to a people. This reconfiguration of his heritage was, in large part, shaped by the many Jews he knew in Berlin who had tried to assimilate into German culture. Most Jews in Germany preferred this approach, which sought to "overcome anti-Semitism by dropping nearly everything Jewish," as Einstein put it. He considered this attempt to blend in—"pussyfooting," he called it—servile and idiotic, and was happy to say so to people's faces.

Assimilation was more common in Western than in Eastern Europe, and Einstein especially disliked the way that in Germany many assimilated Jews viewed themselves as more refined than the mostly unassimilated Jews from countries such as Russia or Poland, and therefore superior. "It was only when, at the age of thirty-five, I got to Berlin that I understood the Jewish community of destiny, and that I felt a duty to oppose, as far as I could, the undignified demeanor of my Jewish colleagues."

He did not rediscover his faith. Judaism, as Einstein conceived it, was not a question of religion. To use a metaphor he employed: a snail may be a creature that occupies a snail shell, but this does not serve as a definition; were the snail to rid itself of its shell, it would still be a snail. He conceived of Judaism, he once wrote, as a "community of tradition." His solidarity with the Jewish people was, in his words, a solidarity with his "tribal companions" rather than religious fellows.

In early 1919, it was to Zionism that Einstein turned as his way of embracing his "tribe." Persuaded in part by the recruitment efforts of the Zionist leader Kurt Blumenfeld, Einstein overcame his instinctive objections to the nationalistic element inherent in the movement—that is, the creation of a Jewish state—and was persuaded that a Jewish home in Palestine would provide Jews with an inner security and freedom they had not yet known.

Walking home with Blumenfeld after one of the latter's lectures, he declared, "I am *against* nationalism, but *for* the Zionist cause. The reason has become clear for me today. If a person has two arms and constantly says, 'I have a right arm,' then he is a chauvinist. If a person however lacks a right arm, then he must do everything to substitute for that missing limb."

Once he had given his support, he never withdrew it. Despite not officially joining any Zionist organization, he often lent his weight in support of the movement's goals, especially the establishment of a Jewish university in Palestine. A Jewish homeland, he believed, would provide "a center of culture for all Jews, a refuge for the most grievously oppressed, a field of action for the best among us, a unifying ideal, and a means of attaining inward health for the Jews of the whole world."

Just as Einstein was committing to this newfound sense of Judaism and to aiding Jews in whatever way he could, so Germany was becoming more openly antisemitic. Since the First World War, in response to crushing reparations imposed by the Allies, a soothing, insidious myth had been propagated in the right-wing press: defeat had come as a result of betrayal at home. The army had been undermined by pacifist, internationalist, and anti-military sentiments on the home front: the civilian population and its leaders had denied it support at a vital moment in the war. This narrative very soon transformed into something simpler, and the blame for the country's humiliation was placed almost entirely on the country's Jews.

This was enough in itself to encourage Einstein to embrace and defend his Jewish feeling. His first public stand against antisemitism came in the summer of 1920, in the form of a personal defense. On August 24, a right-wing nationalistic organization, the Working Party of German Scientists for the Preservation of a Pure Science, held a rally at the Berlin Philharmonic Hall, the purpose of which was to attack the legitimacy of relativity and the character of its creator. Speaking first was Paul Weyland, an engineer who had written several politically minded articles vilifying relativity. He had latched on to the fact that the public, and some scientists, were concerned by the theory's abstract rather than experimental basis, and the way it threatened much "traditional" science with what he saw as its "Jewish nature." Relativity, Weyland declared at the rally, was spurious, a con for publicity, and on top of all that, it was plagiarized. The next speaker was the experimental physicist Ernst Gehrcke, who said effectively the same thing as Weyland, but in scientific language.

Halfway through his speech, a whisper rolled around the hall—"Einstein," the listeners were saying. "Einstein, Einstein." Albert was sitting in one of the boxes for all to see, there to watch the show and mock it in the open. Although he was truthfully enraged by his detractors and their blatant prejudice—and would respond to the meeting a few days later with an article attacking them and refuting their arguments—for now Einstein was all smiles and calm. Along with his friend Walther Nernst, he punctuated the proceedings with loud rounds of laughter and applause. When all was finished, he pronounced the meeting "most amusing."

45

The few, outrageous minutes of a total solar eclipse—when the moon passes in front of the sun, obscuring it—are a neat reminder of the absurdity of the universe. We are able to experience them because of a cosmological fluke. The sun is around four hundred times larger than our moon, and yet it also just happens to be around four hundred times farther away from Earth. The two celestial bodies, therefore, appear to us to be exactly the same size in the sky. At the totality of an eclipse, the black disk of the moon blocks the sun with fascinating tidiness, revealing only the subdued, pearly glow of the sun's outer atmosphere: the corona. Stars appear brighter in the sky.

On May 29, 1919, a rather proper, rather hot Englishman was taking pictures of just such a moment, on the small island of Príncipe in the Gulf of Guinea, off the west coast of equatorial Africa. He barely looked up to see the event for himself, so concerned was he with slotting large photographic plates into the astrograph he'd brought with him, months ago, from Britain. He ignored the mosquitoes and the dampness in the air from an earlier storm. He had a lot of pictures to take in very little time.

During the First World War, Arthur Stanley Eddington was in his mid-thirties. As director of the Cambridge Observatory, he was fast on his way to becoming one of the finest astronomers Britain had ever produced. He was also a devout Quaker and had made it clear that his faith prevented him from fighting in the ongoing war in Europe. In 1917, his colleagues in Cambridge realized that Eddington might well end up in a prison camp as a result of his stance, and managed to secure him an exemption on the grounds that his work was necessary for the war effort. The Home Office sent Eddington the appropriate form to sign, which he did. However, firm in his beliefs, he felt obliged to add a postscript, claiming that were he not deferred on these grounds he would even so refuse to be called up, as a conscientious objector. Not knowing quite what to do, the Home Office retracted the offer.

The Astronomer Royal, Sir Frank Dyson, stepped in and succeeded in reinstating Eddington's exemption. This time it came with a condition: Eddington would have to undertake an important scientific mission. Dyson volunteered him as leader of an expedition. Eddington had always been an enthusiastic advocate for general relativity, making him a complete anomaly in Britain, where the work of German scientists was generally ignored, censored, or derided. Eddington did not subscribe to the spirit of the times. He believed that science had no borders and that in internationalism lay the road to peace. Dyson pointed out to him that a golden opportunity to prove the validity of this strange German theory was just around the corner.

General relativity predicted that massive objects would seem to bend light, as the shortest path for light near, say, a star would be along a curved patch of space-time. The best chance of detecting this deflection of light on Earth was during an eclipse—and one was due in 1919. When it happened, the star cluster Hyades would be behind the sun. The light from each of its stars would fly toward Earth, bending around the sun before reaching us. It would appear, therefore, that the stars had moved position in the sky. Dyson may not have fully understood relativity, but like Eddington, he was aware that it would be an almighty coup for an English astronomer to prove the ideas of a scientist working in the heart of the German capital.

Eclipses are, in fact, quite common: one happens every year or two somewhere over the Earth's surface. The reason that they seem rare, even miraculous, is that on each occasion the shadow of the moon crosses only a small part of the globe. The umbra of the 1919 eclipse would pass over Brazil and the Atlantic Ocean, then dip through the skies of equatorial Africa. There was no point in Eddington sitting around and waiting for it in Cambridge.

With the war still playing out, a shortage in materials, and German U-boats patrolling the seas, Eddington set about planning his expedition. It was decided there should be two teams, in an attempt to beat the chances of bad weather. Eddington and the gifted technician E. T. Cottingham would head to the island of Príncipe, the astronomers Andrew Crommelin and Charles Davidson to the city of Sobral, about seventy miles inland from the northeastern coast of Brazil. Both teams would take measurements and compare them on their return to London.

The physicists separated at Madeira, the Portuguese island near Morocco. The Brazil team continued on, while Eddington and Cottingham—and their astrographic camera—were put ashore. They would have to find their own way from there to Príncipe, but this turned out to be easier said than done. It took nearly a month to hunt down a ship that would take them the rest of the way. Eddington used the dead time to scale the surrounding mountains and to visit Madeira's casino—although not to do any gambling, he assured his mother in his letters home. It was only because they served particularly good tea there.

Once they reached Príncipe they spent their time having huts built, chasing off monkeys, and battling with mosquitoes. One evening, a plantation owner invited them both for supper and placed full bowls of sugar on the table. They were shocked—with the German blockades during the war, neither had seen sugar for nearly five years.

Three weeks after arriving on the island, Eddington, Cottingham, and some local porters hauled their machinery by mule to a plateau far from the central mountains and the threat of storms. Eddington knew the eclipse would occur at precisely five seconds past 2:13 p.m. Cottingham counted him down, and he began. The sky was never fully clear of clouds. Out of sixteen photographs, he ended up with only two that were usable, each showing five blurred stars.

Things hadn't fared much better in Sobral. There, the plump, brooding rainforest clouds had cleared the night before, and the equipment suffered from too much sun rather than a lack of it. The mirror of their astrograph expanded in the heat, changing the focus and spoiling the definition of their images. But Crommelin and Davidson had also taken a telescope with them, as a grudging backup. It took the best pictures of the entire expedition.

What the teams were looking for was the minutest deviation of the stars' positions. Einstein had predicted a deflection of starlight of 1.7 arc seconds—that is, slightly more than 1/2000 of a degree. This is small-scale stuff. To someone on Earth looking at the night sky with the naked eye, the stars would seem to have moved far less than a hair's breadth from their normal positions.

It was on the basis of this tiny change that Einstein's theory, and a new understanding of the nature of the universe, would either stand or fall.

46

The photograph used on the cover of the Berliner Illustrirte Zeitung, *December 14, 1919.*

I t's Christmastime in Cambridge, 1933. In the Senior Combination Room at Trinity College, five men sit around the old fireplace, smoking long clay pipes in accordance with seasonal tradition. They are Ernest Rutherford, the father of nuclear physics; the astronomer and well-known popularizer of science Arthur Stanley Eddington; Maurice Amos, the retired chief judicial adviser to the Egyptian government; the noted geometer Patrick Du Val; and Subrahmanyan Chandrasekhar, an awestruck twenty-three-year-old physicist who, fifty years later, would be awarded the Nobel Prize for his work on stellar evolution.

The conversation continues well past midnight. Amos turns to Rutherford.

"I do not see why Einstein is accorded greater public acclaim than you," he says. "After all, you invented the nuclear model of the atom; and that model provides the basis for all of physical science today, and it is even more universal in its applications than Newton's laws of gravitation. Whereas, granted that Einstein's theory is right—I cannot say otherwise in the presence of Eddington here—Einstein's predictions refer to such minute departures from the Newtonian theory that I do not see what all the fuss is about."

Rutherford, in reply, looks at Eddington. "*You* are responsible for Einstein's fame," he tells him, only semi-jokingly.

It's easy to pinpoint the day Einstein became famous. On November 7, 1919, the London *Times* ran this headline:

REVOLUTION IN SCIENCE
NEW THEORY OF THE UNIVERSE
NEWTONIAN IDEAS OVERTHROWN

The article beneath reported on the joint meeting of the Royal Society and the Royal Astronomical Society, which had taken place the previous afternoon at Burlington House in London. They had gathered solely to announce, and debate, the results of Eddington's expedition—that is to say, to pronounce on the validity of relativity.

Above the great and the good of Britain's scientific establishment hung a portrait of Sir Isaac Newton, looking down sternly at the fellows. J. J. Thomson, the discoverer of the electron and the president of the Royal Society, was chairing the meeting. The Astronomer Royal, Frank Dyson, presented the findings, concluding, "After a careful study of the plates, I am prepared to say that there can be no doubt that they confirm Einstein's prediction."

There were murmurs of reservation. Ludwik Silberstein, the Polish physicist who in 1914 had authored an important textbook on relativity, advised caution. Gesturing to Newton's portrait, he warned, "We owe it to that great man to proceed very carefully." But no one was really listening.

Thomson captured the mood of the room as he brought the discussion to an end: "This is the most important result obtained in connection with the theory of gravitation since Newton's day."

The *Times* article was surrounded by stories on labor disputes, the price of coal, and the defeated Germans, as well as an official press statement from King George V calling for a two-minute silence on Armistice Day, the first anniversary of which fell in a few days' time. The announcement of a new theory of the universe strikes a jarring note of progress among these somber stories. During their late-night conversation at Trinity, Rutherford conjectured that it was the fact that an astronomical prediction by an "enemy" scientist, made during the war, had been verified by one of their own that had really caught the public's imagination.

When Einstein is first mentioned in the *Times* piece, he is described as "the famous physicist, Einstein." While it's true that in the physics community he was at least moderately well known, almost no one else in Britain would have recognized his name. For the general public, Einstein was famous on first meeting. It was all pretty sensationalist stuff. But the press coverage in Britain was positively understated compared with that of the *New York Times*.

The American paper didn't have a science correspondent in London, and so the editors called on the services of their golf expert, the amiable Henry Crouch, who happened to be in the country. Initially, Crouch decided that he wouldn't bother to go to the meeting. When he changed his mind at the last moment, he found that he couldn't get in. No matter: he would simply read the London *Times*, pilfer and embellish what he needed, then file his story as if nothing had happened. However, in the evening—in what must have come as a blow—he was informed that the London *Times* only intended to cover the meeting in brief, deeming it effectively incomprehensible. Ever intrepid, Crouch telephoned Eddington himself for a summary of what he had missed. Unsurprisingly, Crouch failed to understand a word and had to ask for a simpler version that an ordinary newspaper reader could cope with. He reported back, on November 8, unsure if what he had written had any meaning:

ECLIPSE SHOWED GRAVITY VARIATION
Diversion of Light Rays Accepted as Affecting
Newton's Principles.

HAILED AS EPOCHMAKING
British Scientist Calls the Discovery One of the Greatest
of Human Achievements.

The paper, and the readers, were taken with the story. "Epochmaking" hardly seems like restraint, but somehow the editors thought that Crouch had downplayed the whole event and asked him for more. He sent a special cable to the *New York Times* the following day, and this time the headliners went to town:

LIGHTS ALL ASKEW IN THE HEAVENS
Men of Science More or Less Agog Over Results of Eclipse
Observations.

EINSTEIN THEORY TRIUMPHS
Stars Not Where They Seemed or Were Calculated to be, but
Nobody Need Worry.

A BOOK FOR 12 WISE MEN
No More in All the World Could Comprehend It,
Said Einstein When His Daring Publishers Accepted It.

This account is gloriously, exuberantly inaccurate. The stars were exactly where they were supposed to be. Indeed, that was the whole point: Einstein had predicted their position correctly. Crouch's twelve wise men were entirely fictitious, as was the quotation from Einstein. But the incomprehensibility of relativity was a large part of its fascination and, as the golf expert recognized, worth wrapping in myth.

While scientists were perhaps not "agog," many were certainly bemused by the theory and unwittingly contributed to the narrative of Einstein's gift as a seer. After the announcement of the results at Burlington House, as the members were dispersing, Silberstein came up to Eddington.

"You must be one of three persons in the world who understands general relativity," he said.

Eddington demurred, but Silberstein would not have it.

"Don't be modest, Eddington."

"On the contrary," he replied. "I am trying to think who the third person is!"

The *New York Times*' mythmaking did not stop with twelve wise men. In December, they sent a reporter to interview Einstein at his home in Berlin. Here, claimed the resulting piece, Einstein "observed years ago a man dropping from a neighboring roof—luckily on a pile of soft rubbish—and escaping almost without injury." Having landed, the man apparently found an opportunity to tell Albert that when falling he had not experienced the effect of gravity; he had felt weightless. But there was no such man, nor such a pile of rubbish—the reporter had made it up.

In Germany, the press was more taciturn about the whole event. There had been a smattering of straightforward, factual articles published since mid-November, mainly derived from the London *Times* report, but nothing in the way of fanfare. This is hardly surprising: the country was shattered in the aftermath of its defeat in the First World War. As Einstein wrote to Heinrich Zangger, "Here [in Berlin] all conditions are variable and not all for the best; large-scale corruption and impoverishment . . . The drawbacks of the defeat are being immediately felt, the benefits only bit by bit." Winter had set in early and everything was in short supply. Millions of people had barely any fuel or food. Refugees crammed into the city and homelessness soared. Neither lights, gas, nor water were guaranteed. The Einsteins, whose apartment was a large, top-floor affair, were required to rent out one of their rooms.

It wasn't until twelve days after the *New York Times* interview, on December 14, that the *Berliner Illustrirte Zeitung* afforded Einstein the prestige he had gathered so quickly elsewhere. It published a giant portrait photograph of him on the cover. His face almost completely fills the space. His mustache is neat, his hair dark and not yet wild. He is posed in contemplation: his cheek resting on his hand, his eyes downcast, almost sorrowful, wrapped in his thoughts. Beneath the portrait run the words "A new luminary for the history of the world: Albert Einstein, whose research means a total revolution in the way we see the world and whose findings are equal to those of a Copernicus, Kepler and Newton."

Albert celebrated his success by buying a new violin.

47

On April 2, 1921, Einstein arrives in America. Alongside Elsa, his traveling companions consist of a delegation of the World Zionist Organization, as well as its president, Chaim Weizmann, a talented and successful biochemist who will go on to be the first president of Israel. It is at Weizmann's invitation, one could say his insistence, that Einstein has made the trip. The delegation wants to raise funds to help settle Palestine and, in particular, for the creation of a Hebrew University in Jerusalem. As Einstein knows full well, he is accompanying the delegation to act as a propaganda machine for their tour, and to be "shown around like a prize ox."

On board the *Rotterdam*, docked in New York Harbor, Einstein and Weizmann are met by a crowd of reporters. It is cold. Einstein is wearing a thick gray coat and a felt hat, carrying a violin case in one hand and a pipe in the other. Elsa and Albert strike poses for the cameras for half an hour, then he is subjected to a press conference, which he rather enjoys.

"Can you give us a one-sentence summary of relativity?" one reporter asks.

"All my life I have been trying to get it into one book," Einstein replies. "He wants me to state it in one sentence!"

Weizmann is asked whether he has had any luck understanding relativity with the professor's help. "During the crossing, Einstein explained his theory to me every day," he says, "and by the time we arrived, I was fully convinced that he understands it."

There is a tugboat to the shore, a drum and fife band, and then an hours-long motorcade through the Jewish neighborhoods of the Lower East Side, before arrival at the Commodore Hotel shortly before midnight, and thorough exhaustion.

Einstein does not say much at any of the large events on the tour. At one meeting, attended by eight thousand people, he delivers only this address: "Your leader, Dr. Weizmann, has spoken, and he has spoken very well for us all. Follow him and you will do well. That is all I have to say." At their official welcome to the city, at a ceremony at city hall, he says nothing at all. While Weizmann only sees a smattering of polite applause, Einstein is lifted up onto the shoulders of his colleagues and carried to a car through a cheering crowd.

After three weeks in New York, Einstein and a group of delegates visit the White House and meet with President Warren G. Harding. As he poses for photos with Einstein, Harding is asked whether he understands the professor's theory. He smiles and confesses he doesn't understand it in the least.

There is a reception at the National Academy of Sciences at which those being honored give long speeches about their interests and research. There is a paleontologist and an expert on North American birds. Prince Albert of Monaco, a keen oceanographer, speaks about his investigations and his purpose-built yachts. A whole series of hookworm specialists give increasingly tedious speeches.

Einstein leans over to a Dutch diplomat next to him and, beaming, declares quietly, "I just got a new theory of eternity."

Einstein's tour continues to Chicago, Princeton, and New Haven. In his two days in Boston he engages in six receptions, a breakfast with business professionals, a luncheon, and one kosher dinner for five hundred people. In Hartford, he is led through the city by a motorcade of around a hundred cars, headed by motorcycle policemen, a brass band, and a car carrying the four most prominent rabbis of the area, while fifteen thousand spectators line the route. Streets, shops, houses, and cars are decorated in American, Jewish, and—for some obscure reason—English flags. Children present him with flowers.

Most Jewish business owners in Cleveland declare a half day for Einstein's arrival. When they step off the train, he and Weizmann are greeted by thousands of citizens who are being held at bay by a group of retired Jewish veterans in uniform. Yet another motorcade—this time of two hundred cars and led by a military band—winds through the city, taking Einstein and Weizmann to their hotel, but stopping off at the Hebrew School for a meeting with the staff and two thousand children lasting all of ten minutes. People grab at Einstein's car and pull themselves onto the running boards, while the police wrestle them down. Later, they attend a banquet for six hundred guests.

Back in New York, at the end of May, preparing for his departure, Einstein writes to his friend Michele Besso, stating that he is satisfied to have been useful, but that it's been a grueling two months: "It's a miracle that I endured it."

48

Niels Bohr, thirty-four, tall and shy, gave off the impression that he was both ill at ease and happy about it. It was April 1920 and he had come to Berlin from Copenhagen for a series of lectures. He took the opportunity to visit Einstein for the first time. When he arrived at 5 Haberlandstrasse, Bohr presented his host with a hamper containing butter, cheese, and other flavorsome delights. This was much appreciated by both Albert and Elsa, who were still enduring the food shortages of postwar Germany.

In 1913, Bohr had refined the model of the atom and in the process had started a new era in the development of quantum mechanics. He had been working with the physicist Ernest Rutherford in Manchester. Rutherford, who was a fantastic experimentalist, as well as rather good company, had conceived of a new model of the atom in 1911 in which electrons orbited a central nucleus. Bohr had solved a fault in this model. Rutherford had assumed that electrons circled around the nucleus wherever they liked within the atom, at undefined distances from its center, and as a result he had predicted that electrons should lose energy as they orbited, to the point where they eventually collapsed into the nucleus. The problem was that this didn't actually happen. In other words, Rutherford's model of the atom was unstable, whereas in nature atoms are very stable. Bohr's atom, by contrast, contained electrons that could only exist in certain orbits—at certain "energy levels"—and nowhere else. An electron could gain or lose energy only in certain discrete amounts, and if it did gain or lose energy it would "leap" from one energy level to another. An electron could exist *only* at the prescribed energy levels; nothing in between was allowed. Bohr's model represented the stable behavior of atoms perfectly.

Einstein labeled Bohr's work "the highest form of musicality in the sphere of thought"—indeed, he was slightly jealous of it, claiming to at least one scientist that he'd once had a similar idea himself, but hadn't dared publish it. In 1916 he used Bohr's atomic model as the basis for a series of papers exploring the subatomic realm, in the course of which he made an unsettling discovery that would have ramifications for the very notion of reality.

If an atom is bombarded with photons, which is to say particles of light, it will absorb some of them, before it then emits photons itself, gaining and releasing energy. It had been thought that the photons emitted from an atom burst forth in all directions at once, in a sort of ring. Einstein showed that an emitted photon has momentum, that it is emitted in a particular direction. But he also discovered that there was no way to tell which direction the photon would be emitted in, or when. One could calculate the *probability* of a photon taking a certain direction at a certain time, but that was the most one could do. It was all down to chance. The idea that randomness could be an intrinsic part of the universe threatened the notion of cause and effect on which most of physics, including general relativity, was based. More than that, it threatened the idea that the universe could be understood at all.

When Bohr and Einstein sat down on that spring day in Berlin, they embarked on an argument about the importance of probability and causality. While Einstein kicked against the presence of chance in the foundations of the world, Bohr insisted that the only course was to follow the physics and to abandon causality. Neither convinced the other to shift his opinion.

The second time the two men met was in Copenhagen three years later. Einstein was traveling back from Sweden. Bohr came to pick him up from the train station and they boarded a streetcar to make their way to Bohr's home. During their journey, they began to talk so animatedly that they went well past their stop. Finally realizing their mistake, they got off and took a streetcar in the opposite direction, only to fall back into talking and miss their stop again. And on they went, riding back and forth several times before finally reaching their destination.

Bohr and Einstein were never exceptionally frequent correspondents, but whenever a letter arrived or they met each other again they would engross themselves in conversation about the state of quantum mechanics or the true nature of things. So long-standing was their disagreement that when Einstein wrote to Bohr in 1955, not about physics but asking him to sign a public declaration advocating peace in the atomic age, he began his letter, "Don't frown like that!"

Their arguments, however, did nothing to diminish their obvious mutual respect, nor did it make either like the other less. "Einstein was so incredibly sweet," Bohr wrote late in his life. "I want also to say that now, several years after Einstein's death, I can still see Einstein's smile before me, a very special smile, both knowing, humane and friendly."

49

By the time he was awarded a Nobel Prize, Einstein had been nominated sixty-two times, and his nominees had included eight Nobel laureates. It was generally agreed that the award was overdue.

One reason for the delay was that the Nobel Academy was biased against theoretical physics, preferring science that could be tested by experiment. Relativity, it was argued, was unproven, and a little otherworldly. It didn't do much for Einstein's case that he happened to be Jewish. The academy continued to hold out from recognizing relativity even after Eddington's successful solar eclipse expedition. The 1920 prize was given to the Swiss physicist Charles-Edouard Guillaume for his discovery of anomalies in nickel steel, and the 1921 prize was not awarded at all. Better no winner than Einstein.

Einstein's reputation and fame could not be ignored, however. In 1922, he was retrospectively awarded the prize for 1921. Niels Bohr took the prize for 1922. Even so, Einstein wasn't awarded his Nobel Prize for relativity. Instead, it was for his discovery of the law of the photoelectric effect, which had laid out the modern conception of light as a particle. This was worthy of an accolade in its own right, but that the academy should single it out to the exclusion of relativity seemed jarring.

Einstein didn't attend the award ceremony in Sweden. In June 1922, Einstein's friend Walther Rathenau, the German foreign minister, was assassinated by members of an ultranationalist, antisemitic organization. The police advised Einstein that he should perhaps lie low or even leave Berlin for a while, as his name was known to be on target lists held by Nazi sympathizers. Initially, he was not to be shaken, and in August he defiantly agreed to be exhibited in a car, driven around a huge pacifist rally in the city. But a month later, after a fortuitous offer from a Japanese publisher, he decided to undertake an extensive tour of Asia and what is now Israel. The offer from the academy came shortly before he was due to travel, and with the atmosphere as it was in Berlin, it seemed imprudent to cancel the trip.

At the ceremony in December the chairman of the awards committee made it clear in his speech that, while Einstein was best known for relativity, that was not why he had received the Nobel Prize. Relativity, he claimed, essentially had to do with epistemology—the philosophy of knowledge—rather than science. To receive his prize money, Einstein needed to give a Nobel lecture, as was the case for all laureates. And when he was finally able to deliver his talk—in Gothenburg in July 1923, to an audience of two thousand people, including the king of Sweden—he did so exclusively on the "Fundamental Ideas and Problems of the Theory of Relativity."

The money awarded to Einstein was significant—120,000 Swedish kronor—but he didn't benefit from it directly. In his divorce agreement with Marić, he had suggested that if he ever won the Nobel Prize, the money should be paid in full to her and their sons, so that they might live off the interest. Mileva used the money to buy three rental properties in Zurich.

50

Einstein's tour of Asia took nearly six months. He and Elsa set sail in early October 1922, and they arrived back in Germany in March 1923. He kept a travel diary, in which he documented his busy days meeting emissaries and academics, and his excursions to temples, restaurants, and mountains. On the voyage out, he read the psychiatrist Ernst Kretschmer's recent book, *Physique and Character*, which asserted that certain mental disorders were more common in people of certain physical types. Einstein uncritically took the book's thesis to heart, in part because it flattered a tendency he already possessed, to generalize and classify based on limited personal experience.

Near the beginning of the tour, he and Elsa took an early-morning trip through the Hindu quarter of the city of Colombo, the capital of what is now Sri Lanka. As he wrote in his diary:

We drove in individual little carts that were drawn on the double by herculean and yet fine people. I was very much ashamed of myself for being a part of such despicable treatment of human beings but couldn't change anything. These beggars of majestic

proportions descend in droves on any stranger until he has ca-
pitulated before them. They know how to entreat and to beg
until one's heart is wrung out . . . For all their fineness, they give
the impression that the climate prevented them from thinking
back or ahead by more than a quarter of an hour . . . Once you
take a proper look at these people you can hardly appreciate
Europeans anymore, because they are softened and more brutal
and so much rougher and more covetous—and therein unfortunately lies their practical superiority, their ability to take on
big things and carry them out. In this climate, wouldn't we, too,
become like the Indians?

After landing in Hong Kong, he visited the Chinese quarter with
Elsa, and recorded his thoughts about the inhabitants and the area:

Industrious, dirty, numbed people. Houses very uniform, veran-
das arranged like beehive-cells, everything built close together
and monotonous . . . Quiet and civility in all doings. Even the
children are spiritless . . . It would be a pity if these Chinese
pushed out all other races. For the likes of us the mere thought
is unspeakably boring.

The laborers he encountered in Hong Kong were, he wrote, the
"most pitiful of people on Earth, cruelly mistreated and worked to
death in reward for modesty, gentleness and frugality."

The Japanese he found to be "similar to Italians in temperament,
but even more refined, still entirely drenched in their artistic tradition, not nervous, full of humor . . . Earnest respect without a trace of
cynicism or even skepticism is characteristic of Japanese. Pure souls as
nowhere else among people. One has to love and admire this country."
Yet he assessed that the "intellectual demands of this nation seem to

have been weaker than its artistic ones," and wondered whether this was due to a "natural predisposition."

In Jerusalem he walked to the Wailing Wall on the Sabbath, where he saw "dull ethnic brethren, with their faces turned to the wall," who bent "their bodies to and fro in a swaying motion. Pitiful sight of people with a past but without a present. Then diagonally through the (very dirty) city teeming with the most disparate of religions and races, noisy, and orientally alien."

51

While in Tokyo for the Japanese leg of the tour, Einstein and Elsa stayed at the Imperial Hotel, an impressive new building designed by the master American architect Frank Lloyd Wright. The hotel was so new that part of it was still being constructed.

While Einstein was a guest there, a courier came to his room to make a delivery. Einstein wanted to thank the courier, but either he didn't have any change on him or, as was local custom, the courier refused a tip. Not wanting to send the man away empty-handed, he wrote two small notes on the hotel stationery.

On one sheet of paper he scribbled, "A calm and modest life brings more happiness than the pursuit of success combined with constant restlessness." On a second—evidently pushed for both time and ideas—he wrote, "Where there's a will there's a way." These he signed, joking that with a bit of luck they might end up being valuable.

In 2017, at auction in Jerusalem, the notes sold for $1.56 million and $240,000, respectively.

52

The Einstein Tower in Potsdam, Germany.

The Einstein Tower is a solar observatory on the outskirts of Potsdam. It was built with Einstein's active support with the express purpose of finding evidence to either prove or disprove general relativity.

Constructed between 1919 and 1924, the building is a marvel. Designed by Erich Mendelsohn, it is regarded as the most important example of expressionist architecture. It looks a little like a stylized rocket ship blasting off, firing smoke into the ground in undulating clouds. Its telescope is housed in a smooth, curved, cream-colored tower with distinctive angled windows. Cosmic light is fed from the observatory to a partially submerged, elongated laboratory.

There were many technical difficulties during the tower's assembly, in part because it was made from reinforced concrete, something that set it apart from the red and yellow brick structures that surround it on the campus of the Astrophysical Observatory. It was designed, quite deliberately, to be new; to reflect the majesty and mystery of the universe Einstein had exposed. It was inevitable, therefore, that some people didn't like it. One commentator called it "a cross between a New York skyscraper and an Egyptian pyramid." It was not a compliment.

Mendelsohn gave Einstein a tour of his work. The two wandered through the building for some time, Mendelsohn nervously waiting for a sign of approval from the great professor. Einstein said nothing at all. Only hours later, at a meeting of the building committee, did Einstein stand and whisper his judgment in the young architect's ear. It amounted to one word: "Organic."

53

Einstein in 1925.

A WALK TO THE OFFICE, 1925

The morning passes slowly. After a pleasant breakfast, Albert works. At eleven o'clock, one of his students, Esther Polianowski, arrives at the flat. Einstein championed her application to the university a few years ago. Recently she has expressed misgivings about staying in Germany, and so he has invited her over to talk the matter through.

She sits in the study, her eyes wandering to the giant, ornate, dark-wood bookcase, to the imposing telescope, to the globe tucked in the corner. They only just get through pleasantries before Elsa interrupts them with the news that two Orthodox Jews are downstairs, asking to meet her husband. Show them up, of course, he says.

"*Shalom aleichem,*" they say in greeting: "Peace be unto you." As neither Albert nor Elsa knows Hebrew, Polianowski translates. They are visiting Berlin and wish to have the privilege of shaking hands with the great Jew. It prolongs life, they say, to set eyes on a great man. They add that "it is a blessing to set eyes on a king," though Polianowski thinks it best to leave this out.

Having shaken the hand they wished to shake and said what they wished to say, they depart.

There is a pause. "That was quite nice," Elsa says, "but we are never left in peace."

Einstein suddenly wants to know what the date is. "I'd forgotten that there is a meeting of the Academy of Science. Will you be going on to the university this morning?" he asks Polianowski. "Good, then we'll go together. We'll have plenty of time."

As he puts on his tatty overcoat and hat, he overhears Elsa warning Polianowski not to let him walk. "It's bad for his health."

Outside 5 Haberlandstrasse he smiles at the bright day. "What a lovely morning! Let's not go by underground. It won't do me any harm to walk. How are things going?"

Polianowski has no choice but to fall in step with her teacher, who has already set off down the road. "Thank you, not really well. I haven't got any further with the problem you set me. I shall never be a theoretical physicist, I shall never be creative."

Albert avoids the little puddles on the gray pavement because of the holes in his shoes.

"You've picked up quite a lot in a short time," he tells his student.

"Very few women are creative. I should not have sent a daughter of mine to study physics. I'm glad my wife doesn't know any science; my first wife did."

"Madame Curie was creative."

"We spent some holidays with the Curies," he replies. "Madame Curie never heard the birds sing!"

The pair walk through the Tiergarten, the large, wooded park in the center of the city, with its weathered white sculptures and ruins, its classical curiosities. Einstein admits mournfully that he isn't making any progress in his work.

"Relativity is in the past, and what I am trying now doesn't come off."

His voice is calm. Polianowski thinks that his voice is always calm, and sounds as if it comes from a long way off.

"I have little exchange of ideas with other scientists," he explains, "and not much contact with people. Hardly anything ever comes to me from outside, only from inside."

Leaving the park behind them, they walk beneath the Brandenburg Gate, through the crowds of Pariser Platz and onto the Unter den Linden boulevard.

"I want to go to France," Polianowski says, "to know French really well, so that I could read French literature. I would like to find myself in new surroundings."

"I never look for that," comes the reply as they near the palatial exterior of the university. "I like Paris, but I don't want to go there or anywhere. I should not like to learn a new language; I don't like new food or new clothes. I'm not much with people, and I'm not a family man. I want my peace. I want to know how God created this world. I'm not interested in this or that phenomenon, in the spectrum of this or that element. I want to know His thoughts, the rest are details."

54

Hans Albert Einstein, 1937.

Hans Albert's relationship with his father was not without affection or love. Einstein wasn't lying when he once wrote to him, "You have a father who loves you more than anything else and who is constantly thinking of you and caring about you." For the most part, however, it was a difficult kind of love. Both father and son were proud of the other, but they argued and judged and sulked as much as they ever shared ideas and laughed.

Their relationship was never more strained than when, in 1925, Hans Albert announced his decision to get married. While studying at the Zurich Polytechnic, he had fallen in love with Frieda Knecht. She was nine years his senior, acerbic, very intelligent, under five feet tall, and plain. She was, in short, a good deal like his mother.

By this time Marić and Einstein had been divorced for six years, and although the animosity between them had mostly disappeared, their relationship had grown roots of hostility that could not be pulled free. But in objecting to Hans Albert's girlfriend, they found common ground and a common goal. She was, in their eyes, a devious older woman.

Einstein was also particularly concerned about her height, which he thought was evidence of dwarfism in the family, and he believed her mother was mentally unstable (she actually had an overactive thyroid). He feared that what he saw as hereditary defects would be passed down to any grandchildren: "It would be a crime to bring such children into the world," he lamented to Mileva.

Einstein had a theory that Hans Albert was so smitten because he was inhibited and inexperienced with women: "*She* was the first to grab hold of you, and you now view her as the embodiment of all femininity." He decided that what his son needed to get over the undesirable Frieda was another woman. At one point he even singled out a "good-looking woman in her forties" who might do the job admirably. This idea fell through, but Einstein and Marić did not relent in their opposition. They openly voiced their opinions that the crazy marriage would be a disaster.

"If you ever feel the need to separate yourself from her, *don't let pride alienate you from me*," Einstein wrote to him. "For that day *will* come." He said that he wished to protect Hans Albert from the fate that had befallen him and that had so ruined his son's childhood.

But the more his parents opposed the marriage, the more Hans Albert dug in. He married Frieda on May 7, 1927. Before the ceremony, Einstein advised his son that it would be better if the thing were canceled. That way, the bother and trouble of a divorce could be avoided when the inevitable separation happened. He also advised him not to have children for the same reason—separation would be easier without them.

Einstein still had reservations when his first grandchild, Bernhard Caesar Einstein, was born in 1930. "I don't understand it," he told Hans Albert. "I don't think you're my son."

Eventually, however, he came to accept, and admit, that Hans Albert's marriage was a good one and that it made his son happy. Frieda and Hans Albert were married for thirty-one years, until her death. Bernhard was a healthy child. Indeed, he was soon a favorite of his grandfather. It was he who inherited Einstein's violin.

55

Einstein greeting children, with Gustav Struve
and Georgia Chobe, 1932.

Over the years, Einstein received a lot of letters from children. "I am a little girl of six," one announced in large letters drawn haphazardly across the full width of the writing paper. "I saw your picture in the paper. I think you ought to have a haircut, so you can look better." Having given her advice, the girl, with model formality, signed it, "Cordially yours, Ann."

"I have a problem I would like solved," wrote Anna Louise of Falls Church, Virginia. "I would like to know how color gets into a bird's feather." Dear Mr. Einstein was asked the age of Earth and whether life could exist without the sun (to which he replied that it very much could not). One child asked him whether all geniuses were bound to go insane. Frank, from Bristol, Pennsylvania, asked what was beyond the sky—"My mother said you could tell me."

Kenneth, from Asheboro, North Carolina, was more philosophical: "We would like to know, if nobody is around and a tree falls, would there be a sound, and why." Similarly, Peter, from Chelsea, Massachusetts, drove straight to the heart of human inquiry: "I would appreciate it very much if you could tell me what Time is, what the soul is, and what the heavens are."

Other questions were not quite so fraught. A boy named John informed Einstein that "my father and I are going to build a rocket and go to Mars or Venus. We hope you will go too. We want you to go because we need a good scientist and someone who can guide a rocket good."

Occasionally, skeptical correspondents emerged, such as June, a twelve-year-old student from Trail Junior High School in British Columbia, Canada. "Dear Mr. Einstein," she wrote. "I am writing to you to find out if you really exist. You may think this very strange, but some pupils in our class thought that you were a comic strip character."

In a similar vein, Myfanwy from South Africa had thought Einstein dead:

> I probably would have written ages ago, only I was not aware that you were still alive. I am not interested in history, and I thought you had lived in the 18th c., or somewhere around that time. I must have been mixing you up with Sir Isaac Newton or someone. Anyway, I discovered during Math one day that the mistress . . . was talking about the most brilliant scientists. She mentioned that you were in America, and when I asked whether you were buried there, and not in England, she said, Well, you were not dead yet. I was so excited when I heard that, that I all but got a Math detention!

Myfanwy proceeded to tell Einstein how she and her friend Pat Wilson would sneak around the school at night to carry out astronomical observations, and about her love of science. "How can Space go on forever?" she wondered. "I am sorry that you have become an American citizen," she finished. "I would much prefer you in England." Einstein was obviously taken with Myfanwy's exuberance, as he sent her a reply in which he praised her nighttime escapades and apologized for remaining alive. ("There *will* be a remedy for this, however.")

On his seventy-sixth birthday, Einstein was sent a pair of cuff links and a tie by the fifth-grade children of Farmingdale Elementary School in Pleasant Plains, Illinois. "Your gift," he wrote to them, "will be an appropriate suggestion to be a little more elegant in the future than hitherto. Because neckties and cuffs exist for me only as remote memories."

This was one of Einstein's last letters. He died around three weeks after writing it.

56

In December 1925, the young Austrian physicist Erwin Schrödinger
was holed up in the village of Arosa, Switzerland, with one of his mis-
tresses. He was there for his health: suspecting a mild case of tuberculosis, his doctors had ordered him to rest at high altitude. There, among
the calm of the mountains and deep snow, placing a pearl in each of his
ears when he wanted quiet, he developed a theory that became known
as "wave mechanics."

Schrödinger's theory was inspired by the ideas of Louis de Broglie,
a physicist who in his doctoral thesis of 1924 had showed how to calculate the wavelength of a particle based on its momentum. In 1905,
Einstein had demonstrated that waves can act like particles. What de
Broglie argued was that particles can act like waves.

Wave mechanics provided a set of equations that prescribed how
wavelike particles could behave. On first encounter with the theory,
Einstein and many others were impressed and pleased with its useful-
ness, but it was soon noticed that some implications of Schrödinger's
mechanics were a little problematic. For one thing, the theory stated
that the waves it described would, given time, propagate over a very

large area, much like a ripple on the surface of a lake spreading out and out, making for the shore. But Schrödinger's waves were, of course, also particles—they were electrons and other subatomic objects. To Einstein it seemed almost nonsensical to say that an electron would propagate over such enormous distances. It simply didn't accord with reality.

So Schrödinger's mathematical description of waves raised a question. If it didn't represent literal waves, waves in the real world, what did it represent? Einstein's good friend Max Born, a professor at the University of Göttingen, devised an answer: it represented the *probability* of a particle's location. Which is to say that each particle has what's called a "wave function," and one can use this to predict the likelihood of finding a particular particle in a particular place.

Put an electron in a box. According to this idea, the electron has a number of potential locations spread throughout the box, and it exists in a kind of muddled-up mixture of all these possible positions. This mixture is mathematically represented by the electron's wave function, which gives us the various different probabilities of detecting the electron at the various different locations within the box.

Einstein, consistently throughout his career, was unhappy with quantum mechanics' reliance on probability. In fact, he did not like it at all. He strongly believed, even though evidence suggested otherwise, that at a deep level the universe was not run on chance and that the order apparent in the observable universe was built on order in the subatomic realm.

When debating with the theory's various advocates, he would often tell them, "God does not play dice." To which Niels Bohr had a rejoinder: "It cannot be for us to tell God how he is to run the world"—or in other words, "Einstein, stop telling God what to do."

57

In the summer of 1925, when he was twenty-three years old, Werner Heisenberg traveled to the tiny island of Heligoland in the North Sea, hoping that its beaches and sheer cliffs would allay his bad hay fever. There, in one intense night, he finalized his interpretation of the difficulties of the quantum realm. Heisenberg worked from the premise that he could completely ignore what could not be observed, measured, or proved to be true. This sounds quite reasonable, but in this instance it meant that, in order to develop his theory of the laws that govern the behavior of electrons, he made no effort to describe, or really even to think about, the motions or orbits of electrons, as they could not be observed. Instead, he looked at the light emitted by electrons under different circumstances. If you bombard an atom with light or disturb it in other ways, an electron will produce light. Heisenberg looked at what went in and what came out, and didn't concern himself about what happened in between. The result was a paper so mathematically complicated that he couldn't fully understand it himself. He gave the paper to his supervisor, Max Born, and then went camping, hoping that

Born might be able to figure it out for him. Born did just that, and had the paper published.

Einstein didn't like Heisenberg's approach any more than he liked Schrödinger's wave mechanics. He called it "a big quantum egg" and declared outright to one of his friends that he didn't believe in it. The problem, as far as Einstein was concerned, was that Heisenberg had skipped over the need to actually *understand* what was happening. The mathematics didn't really require you to "know" anything about what the electrons were up to between the input and output—they could be doing anything, and it wouldn't affect Heisenberg's theory. To Einstein that wasn't a good enough description of reality.

In 1926, Heisenberg came to Berlin to give a lecture. Einstein, who had already exchanged a few letters with the radical young man, invited him to visit his house, where they soon fell to arguing, as was only to be expected. Heisenberg thought that he would be able to win his host around to his way of thinking, precisely because it had once been Einstein's way of thinking. With relativity, Einstein had done away with seemingly logical but—crucially—unobservable concepts, such as the ether or Newton's absolute space and time, and produced a sweeping, progressive theory. Heisenberg felt he was up to much the same thing.

"We cannot observe electron orbits inside the atom. A good theory must be based on directly observable magnitudes," Heisenberg insisted.

"But you don't seriously believe that none but observable magnitudes must go into a physical theory?"

"Isn't that precisely what you have done with relativity?"

"Possibly I did use this kind of reasoning, but it is nonsense all the same."

Einstein was at least consistent in his contrariness to his old beliefs. To his friend Philipp Frank he made a similar complaint.

"A new fashion has arisen in physics," he rumbled, "which declares that certain things cannot be observed and therefore should not be ascribed reality."

"But the fashion you speak of was invented by you in 1905!" Frank reminded him with amused disbelief.

"A good joke should not be repeated too often."

58

C ount Harry Kessler, a well-liked patron of modern art, is holding
a dinner party at his Berlin home on February 15, 1926. Albert
and Elsa are in attendance.

"Einstein sublimely dignified," Kessler writes in his diary, "despite
his excessive modesty and wearing laced boots with a dinner jacket.
He has become a little stouter, but his eyes still sparkle with almost
childlike radiance and twinkling mischief."

The other guests include a French diplomat, a newspaper editor, a
playwright, and a countess. Alongside Aline Mayrisch de Saint-Hubert,
the founder of the Luxembourg Red Cross, and the memoirist Helene
von Nostitz—who has inspired work by Auguste Rodin and Rainer
Maria Rilke—there's Gustav Hertz, the nephew of Heinrich Hertz, the
influential physicist whose work Albert had studied while at university.

"Your uncle wrote a great book," Einstein assures him at the dinner
table. "Everything in it was wrong, but it was nevertheless a great book."

Kessler talks to Elsa, who tells him, "Recently, after numerous ad-
monitions, Einstein at last went to the foreign ministry and fetched
the two gold medals awarded him by the Royal Society and Royal

Astronomical Society." Elsa says she met her husband afterward and asked him what the medals looked like, but he didn't know as he hadn't even undone the package. "He has no interest in such trifles."

She supplies the count with another example of his lack of concern for such things. The American Barnard Medal, awarded every five years to outstanding scientists, had just been given to Niels Bohr. Reporting the event, the newspapers recalled that in 1920 it had gone to Einstein. Reading this, Albert showed Elsa the paper and asked, "Is that true?" He had completely forgotten.

And, she continues, Einstein simply won't wear his Pour le Mérite, a prestigious German award. She reports a recent session of the Prussian Academy of Sciences at which the chemist Walther Nernst drew his attention to the fact that it was missing.

"I suppose your wife forgot to lay it out for you. Improperly dressed."

"She didn't forget. No, she didn't forget; I didn't want to put it on."

59

Einstein and Niels Bohr at the 1930 Solvay Conference.

Things have changed, says Niels Bohr in his opening presentation for the Solvay Conference of 1927. For thousands of years, physics has been dedicated to the search for truth, for the incontrovertible underlying *reality* of the universe. No more. Nature is, at its most basic, unknowable. In the subatomic realm—the world of quantum mechanics—both causality and certainty disappear. There is no absolute, complete truth to aim toward.

Bohr's grand summary of the state of physics took into account a recent development. Earlier in 1927, Werner Heisenberg had developed his uncertainty principle, which effectively states that it's impossible to know everything about a particle at once. Some quantum properties—such as momentum and position, time and energy—are linked in such a way that the more that is known about one property, the less is known about the other. The more that is known about position, for example, the less will be known about momentum. If one knows a particle's position exactly, then one will know *nothing about its momentum*.

Albert said very little during the formal presentations of the conference. But away from the main discussions, over meals and on walks, Einstein would attempt to poke holes in this new physics. "One can't make a theory out of a lot of 'maybes,'" Wolfgang Pauli remembered him saying. "Deep down it is wrong, even if it is empirically and logically right."

Einstein made sure to be constructive in his criticism, even when he hoped it might prove destructive. He liked to present Bohr and his team of younger colleagues with a thought experiment. He would imagine, say, a complicated contraption that, theoretically if not practically, could exist, and that could measure all there was to know about a moving particle, with certainty. How, he would ask, does that square itself with quantum theory?

Bohr took Einstein's criticisms very seriously. He would mutter away and talk things through with his colleagues, and by dinnertime they would usually have a solution ready to dismantle Einstein's problem. At times he could hardly sleep for wrestling with Einstein's objections. The physicist Paul Ehrenfest, a friend of both men, got the worst of his worries. "Every night," Ehrenfest reported to his students, "Bohr came to my room at 1 a.m. to say 'just one single word' to me, until 3 a.m." Over breakfast the next day, Albert, acknowledging his earlier defeat, would present another problem he'd thought up, always more difficult than the last. And Bohr would stalk away, muttering to himself once more.

At the next Solvay Conference, in 1930, Einstein came prepared with a particularly devilish and sophisticated thought experiment. Imagine a box filled with a cloud of photons resting on a very sensitive scale. The box is also hooked up to an incredibly accurate clock and has a tiny shutter built into one of its sides, controlled by this clock. At a particular time, the shutter opens and closes so as to let out just one photon. Because of the extreme accuracy of the clock, we can tell the *exact* moment the photon left the box. Added to this, because the box is resting on a scale, its weight both before and after the release of the photon is known, and so the exact mass of the photon can be deduced. As $E = mc^2$ tells us, to know the mass of something allows us also to know its energy. This was an instance, Einstein said, in which we can know both the exact energy a photon carries and the exact time at which it does so—something that was in strict contradiction to the uncertainty principle.

Bohr was stunned. At the university club, he wandered between people, attempting to convince them, for the sake of physics, that Einstein *must* be wrong. But he could not think of an answer to the problem. Einstein and Bohr left the club together, one attendee re-membered, "Einstein, a majestic figure, walking calmly with a faint ironic smile, and Bohr trotting by his side, extremely upset."

Indeed, Bohr couldn't sleep that night. But, by the morning, he had found an answer. Einstein, he'd realized, had not taken into ac-count the general theory of relativity. After the photon escaped via the shutter, the box would become lighter by the weight of one pro-ton. The scale measuring the weight of the box would then rise the smallest fraction higher in Earth's gravitational field. And relativity states that time operates at different rates at different points in a gravitational field. In other words, the small rise of the scale meant that, in fact, one *couldn't* be certain about the time the photon es-caped.

Einstein, to his credit, helped Bohr over breakfast with his calculations to conclude that the uncertainty inherent in weighing the box matched what was predicted by Heisenberg's principle exactly. Bohr was very polite about it all, but it was clear enough to both of them that he'd won the argument.

60

Helen Dukas and Einstein in Einstein's study, Princeton,
New Jersey, 1940.

A tall woman with large, dark eyes and short, dark hair entered the room. She was shy and unassuming, yet there was a severity in her bearing and force in what she did. From the bed, Albert Einstein stretched a hand toward her and, smiling, introduced himself: "Here lies an old corpse." Such were the first words he said to Helen Dukas in April 1928.

While staying at the small resort of Zuoz, in Switzerland, Einstein had managed to incapacitate himself by lifting a large suitcase. He had paid so little heed to his body for so many years that this simple effort brought about a rapid and overwhelming physical collapse. He was diagnosed as suffering from a large heart and had to remain bedridden for four months after returning to Berlin. As he was unable to keep up with his work and correspondence, Elsa decided he needed an assistant.

Dukas agreed to take the job only after being assured that it did not require her to understand physics. In the not uncommon event of someone asking her to explain relativity, she would sometimes provide an answer that Einstein had cooked up especially for her: "An hour sitting with a pretty girl," she would say, "passes like a minute; but a minute sitting on a hot stove seems like an hour—that's relativity."

Although she claimed that she never quite got over her nervousness around Einstein, Dukas was soon treated as a member of the family. She moved to America with her employer in 1933 and, after the death of Elsa, became his housekeeper. She was mistaken by at least one visitor as Einstein's wife. Hans Albert speculated that Dukas was having an affair with his father, but there is no evidence to support this. For the most part, Einstein seems to have paid his assistant as much attention as he might have paid a table—a table of which he was particularly fond.

Dukas worked hard and became doggedly, fervently loyal to her employer, as well as very protective of him, especially when dealing with pesky biographers or anyone else prying into his personal life. She would often run out of the house and shoo the press away, or shout at Einstein not to say anything to them. She was such a fierce guard that, even among the residents of Princeton, New Jersey, she was occasionally referred to as Einstein's personal Cerberus—the three-headed hound of Hades who guarded the entrance to the underworld in Greek mythology.

"That is Miss Dukas, my faithful helper," Einstein once said to a friend. "Without her nobody would know I was still alive."

By the time the family had settled in Princeton, it was Dukas who decided what letters Einstein read and who she thought was worth his time. In Einstein's later years, she would even assess whether or not he needed to hear from his family. Einstein's granddaughter, Evelyn, recalled that more than once she wrote to her grandfather, only for the letter never to reach him. Having decided that Professor Einstein was too busy, Dukas would often read and reply to such letters herself.

In his will, Einstein left his devoted assistant $20,000, the same amount he left his stepdaughter Margot. He gave Eduard $15,000 and Hans Albert only half what Dukas had received. Dukas was also given joint charge over his entire literary estate, which meant that she, along with Einstein's equally austere friend Otto Nathan, had ownership over every word he had ever written, even his letters to his sons. She used this position to control the image of Einstein that was presented to the world. Anything that painted Einstein as less than either a mystery or a secular saint was to be suppressed. For nearly thirty years after Einstein's death, Dukas sought to stop any of his negative aspects from becoming public, going so far as to oversee the work of biographers and deny them access to materials.

Any mention of Einstein's first family was intrinsically a stain on his character. Dukas disliked Mileva Marić intensely, and her animosity continued even after Marić died. When Hans Albert's wife, Frieda, wished to publish her own biography of Einstein, which made use of letters to Marić and his two sons to present a more human side of her father-in-law, Dukas went to court. She made sure the book never appeared.

61

For Einstein's fiftieth birthday, in March 1929, the Berlin city government decided to give him a present: a house in the country. Having recently purchased the Neukladow estate—once the home of Otto von Bismarck's mother—they gave Einstein the lifelong right to reside in a classical-style house on grounds that overlooked part of the river Havel, where he could sail and calculate and think, surrounded by peace and beauty.

When Elsa took a trip to inspect the new vacation home, she was surprised to encounter the previous owners of the property, an aristocratic couple who were not only still living in the house but who also told her that she had no business being on the estate. They were, as it turned out, absolutely correct. For some inexplicable reason, the contract the city government had signed with the couple gave them the right to continue living there. The city owned the land and the house, but it couldn't evict the previous owners.

The embarrassed city tried offering Einstein a large portion of the estate, on which he could build his own house. But this also broke the terms of the contract. A different property was offered, but it was infested with midges and flies, had no water, and was stuck behind some stables. A succession of properties were considered, but all were equally unsuitable. The newspapers gleefully reported on the incompetence of local government.

It was agreed, eventually, that Einstein should simply find his own land and the city would pay for it. Albert settled on somewhere easily enough: a small plot owned by some friends, on the edge of a pretty village called Caputh, at the meeting point of many lakes and flanked by a large forest. The mayor of Berlin asked the assembly of city deputies to approve the spending of twenty thousand marks to buy the land, so they might finally have a present to give the professor, and save face. Einstein hired an architect to draw up some plans and he and Elsa quickly fell in love with the idea of their summerhouse.

Yet another problem arose. The right-wing German nationalists in the city parliament objected to the spending of the money, delayed the vote, and insisted that the subject be put up for a full debate. It was reasonably clear the debate was to be about Einstein himself as much as anything else. When he learned of this, Einstein sent the mayor a letter.

"Life is short, while the authorities move slowly," he wrote. "My birthday is already past and I decline the gift."

62

I n 1929, Einstein received a telegram from Rabbi Herbert Goldstein in New York which read in full "Do you believe in God? Prepaid reply fifty words."

"No one," Einstein laughed, "except an American, could think of sending a man a telegram asking him 'Do you believe in God?'" Even so, despite his amusement, he used twenty-nine of his fifty words to supply a sincere answer: "I believe in Spinoza's God, who reveals Himself in the orderly harmony of what exists, not in a God who concerns Himself with the fates and actions of human beings."

Einstein had read the work of the seventeenth-century Dutch philosopher Baruch Spinoza with his friends at the Olympia Academy during his time in Bern, and he felt a kinship with the ideas he encountered. Spinoza would become his favorite thinker and the guiding star for his own system of belief.

Spinoza didn't believe in the tenets of traditional religion: there is no afterlife, he said, and man is not special. He argued that the Bible was not divinely inspired. Nor did he believe in a traditional God. For him, God doesn't judge our actions, hear our prayers, punish us for transgressions, or reward us for virtue. "Neither intellect nor will pertain to God's nature," he wrote, considering these traits to be human projections.

Einstein's take on this was that "I cannot conceive of a God who rewards and punishes his creatures, or has a will of the kind that we experience in ourselves. Neither can I nor would I want to conceive of an individual that survives his physical death; let feeble souls, from fear or absurd egoism, cherish such thoughts."

Spinoza argued that God does not stand outside of nature. Instead, effectively, He *is* nature—which one may think of as existence itself, the universe and its laws. "Whatsoever is, is in God, and without God nothing can be, or be conceived," he wrote. Or, as Einstein put it: "We followers of Spinoza see our God in the wonderful order and lawfulness of all that exists and in its soul as it reveals itself in man and animal."

Spinoza believed that life is dictated by God, which is the same as saying it is dictated by the laws of nature. Life, then, is deterministic. People do not have alternatives to their actions; they simply act in accordance with the inexorable laws of the universe. "I do not believe in free will," Einstein said in 1932.

On another occasion, Einstein reasoned that the moon, if it were gifted with a sense of self, might think that as it circles Earth it is doing so of its own volition. As we would smile to hear such a thing, so a more intelligent being than us would smile as we proclaimed to act of our own free will. Proper devotion to God, both Spinoza and Einstein believed, comes from trying to understand the workings of the world and then accepting them, so as to exist in harmony with the universe, with God.

In an interview in 1930, Einstein was pushed further to try to define his religion. He responded, appropriately enough, with a parable.

We are in the position of a little child entering a huge library filled with books in many languages. The child knows someone must have written those books. It does not know how. It does not understand the languages in which they are written. The child dimly suspects a mysterious order in the arrangement of the books, but doesn't know what it is. That, it seems to me, is the attitude of even the most intelligent human towards God. We see the universe marvellously arranged and obeying certain laws but only dimly understand these laws. Our limited minds grasp the mysterious force that moves the constellations.

63

Shortly after publishing his equations of general relativity in 1915, Einstein began work on what would become the focus of the rest of his life. He sought to bring together the theory of gravity and the theory of electromagnetic forces, so that they formed a "unified field theory"—a theory of everything.

With his growing fame, his efforts frequently drew unprecedented attention in the press. In late 1928, he submitted a mathematical paper to the Prussian Academy of Sciences. This prompted newspaper reports on Einstein's amazing new theory, even though nobody had yet seen the paper. As journalists from around the world surrounded his building in Berlin, he had to go into hiding at his doctor's villa. And this fevered, clamorous interest only increased when the academy published the work, at the end of January 1929.

A thousand copies were printed, only to sell out immediately, and three thousand more were ordered. One American newspaper printed the entire paper in its pages, along with a far more entertaining companion piece about how difficult it had been to send the Greek letters in the equations via telegraph. In London, the five pages of the paper were pasted side by side in the window of Selfridges so that people on the street could read them. Crowds jostled as they struggled to get close enough to interpret the mass of complicated equations.

From his hideaway, Einstein wrote an article about his new work, which appeared as a special feature in the *New York Times*, and he gave interviews to international magazines. To the *Daily Chronicle* he declared, "Now, but only now, we know that the force that moves electrons in their ellipses about the nuclei of atoms is the same force that moves our earth in its annual course around the sun." This, as it turned out, was nonsense.

In fact, scientifically speaking, Einstein's paper was not that impressive. And it was only his most recent attempt at unifying the forces. By 1929, he already had a long list of failed efforts. During the latest excitement, Wolfgang Pauli, the young and radical quantum physicist, caustically predicted that Einstein would give up this way of thinking within the year. He was almost right. Just over eighteen months later, Einstein had changed direction once again. "So," he wrote to Pauli, "you were right, you rascal."

This cycle continued for more than twenty years: Einstein would work on one method, declaring it the "definitive solution," then within a short time he would abandon the idea and move on to something else. As he had made his name early in life, he could afford to expend his energy chasing something that someone younger, thinking of their career, might not, for fear of failure and wasting time. Being as free as he was, Einstein felt it was his obligation to search where others would not.

But his pursuit of a unified field theory began to isolate him from his colleagues. His work became ever more abstract and mathematical, disconnected from ideas grounded in observable reality. He no longer kept up with the most recent developments in physics, and his work suffered.

He was well aware of his failures. As he wrote to Maurice Solovine in 1948, "I shall never ever solve it."

64

From 1930 to 1932, Einstein visited Oxford for a month each year, at the invitation of Christ Church, a college hoping to secure him for a more permanent position.

Oxford's opulence and formality did not sit well with Einstein. When dining with the senior members of the college at High Table, he would often occupy himself by scribbling notes on a pad hidden on his lap, for fear of being caught. In the main, though, he enjoyed his time there. His Jewishness, his Germanness, and his reputation were accepted and disregarded, and, in a place at home with eccentrics, his own eccentricities were considered unimportant. For the most part, he was taken on his own terms.

On one occasion, Gilbert Murray, the public intellectual and classical scholar, was passing through Tom Quad, the college's large, formal parcel of mown grass and gravel walkways, and caught sight of a smiling Einstein sitting alone with a faraway look. Murray asked him what was on his mind.

"I am thinking that, after all, this is a very small star."

During one of his stays at Oxford he was visited by his mistress

Ethel Michanowski, a Berlin socialite and friend of his stepdaughter Margot. She stayed at a nearby hotel. When Elsa found out, Einstein was blasé:

> Your dismay towards Frau M. is totally groundless because she behaved completely according to the best Jewish-Christian morality. Here is the proof: 1) What one enjoys and doesn't harm others, one should do. 2) What one doesn't enjoy and will only aggravate others, one should not do. Because of #1, she came with me, and because of #2 she didn't tell you anything about it. Isn't that impeccable behavior?

In the end, it was Einstein who grew irritated by Michanowski's presence. "Her chasing me is getting out of control," he wrote to Margot. "I don't care what people are saying about me, but for [Elsa] and for Frau M., it is better that not every Tom, Dick and Harry gossip about it." During her stay, Michanowski sent an expensive gift to Einstein at Christ Church, a gesture he did not appreciate. "The small package really angered me," he wrote to her. "You have to stop sending me presents incessantly." He signed off his letter "with an absolutely scathing look."

Einstein told Margot that the women he was carrying on with meant little to him. Out of all of them, he said, he was only really attached to "Frau L." This was Margarete Lebach, a married Austrian with blond, unruly, bobbed hair. Lebach and Einstein's relationship was not a clandestine one. Whenever she visited the summerhouse in Caputh to go sailing with him, around once a week, she made sure to bring pastries for Elsa. In turn, Elsa would do her best to be absent whenever she knew "the Austrian" was coming, often leaving for Berlin in tears.

Once, a piece of Lebach's clothing was found in Einstein's boat (it is not said which item of clothing) and this discovery led to a fierce family argument, in which Margot pressed her mother to have Einstein end the relationship. But Elsa knew that Albert would simply refuse to do so. He had made it known to Elsa that he believed that people were not naturally monogamous, and that the concepts of emotional and physical faithfulness were societal constructs, falsities born of decorum and correctitude.

Elsa decided that it was worth preserving her marriage. "You have to see him all of one piece," she once said. "God has given him so much nobility, and I find him wonderful, although life with him is exhausting and complicated."

65

Spring was slowly becoming summer during Einstein's stay in Oxford. The quince trees were in blossom, but the cherry had been and gone. It was too late for daffodils, but too early for meadowsweet. The grass was high and the air felt light.

Einstein went for a walk in Magdalen Deer Park, a large meadow crossed with paths and cut up by the river Cherwell, which at some points runs no bigger than a stream.

On one of the bridges stood an undergraduate, nineteen years old and sturdy-looking, who was staring idly at the water. Einstein paused to stand near him. The young man was William Golding, the future author of *Lord of the Flies* and recipient of the Nobel Prize in Literature. At the time, he was studying natural sciences.

Golding was eager to express to Einstein what an honor it was to meet him. Unfortunately, he spoke German as well as Einstein spoke English at that point, which is to say almost not at all. So he smiled and hoped it would do the trick. Silence ensued for five minutes, Golding beaming away all the while, before Einstein decided that something ought to be said.

"*Fisch*," he ventured, pointing to a trout in the stream below.

Golding sought for some way to show that he, too, appreciated reason and science. "*Fisch*," he agreed. "*Ja. Ja.*"

They stood benignly side by side for five more minutes. Then Einstein, the picture of friendliness, drifted away to continue his walk.

66

Einstein didn't have much time for psychoanalysis, regarding it as a questionable, even fraudulent, science. He knew Sigmund Freud a little and liked him, having met him for dinner in Berlin in 1927, but while he was civil to the psychoanalyst, he was certainly not convinced by him. When Freud wrote him a letter for his fiftieth birthday, congratulating him on being happy and fortunate, Einstein sent an oddly pugnacious reply: "Why do you emphasize my good fortune? You, since you have slipped into the shoes of so many people, indeed of humanity in general, hardly had an opportunity to slip into mine."

In 1932, the International Institute of Intellectual Cooperation (the predecessor of UNESCO) asked Einstein if he would like to exchange letters with a recipient of his choice, to discuss a subject relating to politics and war. Einstein chose Freud as his correspondent. His question was a broad one: "Is there any way of delivering mankind from the menace of war?" He went on to lay out his own thoughts on a possible answer, which involved the setting-up, by international consent, of "a legislative and judicial body to settle every conflict arising between nations." All countries would have to accept and abide by the orders of this body. He was, of course, aware that such a system had its drawbacks. "Here, at the outset, I come up against a difficulty," he admitted: "law and might inevitably go hand in hand."

That no such body had been established in the aftermath of the First World War Einstein attributed to "strong psychological factors" that were paralyzing well-intentioned efforts. In particular, he blamed the "craving for power which characterizes the governing class in every nation," and that was "hostile to any limitation of the national sovereignty." How was it possible, he asked Freud, "for this small clique to bend the will of the majority, who stand to lose and suffer by a state of war, to the service of their ambitions?"

They had the schools, the press, and usually the Church under their control, it was true, yet Einstein wondered how even this pervasive influence succeeded in encouraging people to support war so fervently, to the point of sacrificing their own lives. "Only one answer is possible. Because man has within him a lust for hatred and destruction. In normal times this passion exists in a latent state. It emerges only in unusual circumstances, but it is a comparatively easy task to call it into play and raise it to the power of a collective psychosis."

Freud's reply was long and intricate, involving as its starting point a broad-brush history of the development of human society. For the most part, he agreed with Einstein and struck the same pessimistic tone. To end war, he said, a central control that would have "the last word in every conflict of interests" was needed. This body would be impotent without an executive force at its disposal, but Freud saw the establishment of such a force as a forlorn hope. He went on to address Einstein's attempt at psychoanalysis.

You are amazed that it is so easy to infect men with the war-fever, and you surmise that man has in him an active instinct for hatred and destruction, amenable to such stimulations. I entirely agree with you . . .

We assume that human instincts are of two kinds: those that conserve and unify, which we call "erotic" (in the meaning Plato gives to Eros in his *Symposium*), or else "sexual" (explicitly extending the popular connotations of "sex"); and, secondly, the instincts to destroy and kill, which we assimilate as the aggressive or destructive instincts. These are, as you perceive, the well-known opposites, Love and Hate, transformed into theoretical entities . . .

When a nation is summoned to engage in war, a whole gamut of human motives may respond to this appeal; high and low motives, some openly avowed, others slurred over. The lust for aggression and destruction is certainly included; the innumerable cruelties of history and man's daily life confirm its prevalence and strength. The stimulation of these destructive impulses by appeals to idealism and the erotic instinct naturally facilitates their release.

Although there was "no likelihood of our being able to suppress humanity's aggressive tendencies," Freud drew some hope from his analysis. From this "mythology," he wrote, it was easy to work out an indirect way to eliminate war: "If the propensity for war be due to the destructive instinct, we have always its counter-agent, Eros, to our hand." In other words, "All that produces ties of sentiment between man and man must serve us as war's antidote."

The cultural development of humanity, he reasoned, worked against our disposition for war. With the growth of civilization, more people would turn pacifist. Overall, however, his answer to Einstein's question was that, no, there is no way of delivering mankind from the menace of war. Freud joked that their letters would win neither of them the Nobel Peace Prize.

In any case, their exchange soon became outdated and academic, made irrelevant by the movements of history. By the time it was published, in 1933, Hitler had taken power in Germany.

67

The FBI's formal interest in Einstein began in 1932, with the receipt of a letter from the aptly named Mrs. Randolph Frothingham, president of the Woman Patriot Corporation, a group that sought to guard America against "undesirable aliens," especially pacifists, socialists, and feminists. The group had already been angered over Einstein's recent appointment to the new Institute for Advanced Study in Princeton, which he would take up the following year. Now Frothingham's sixteen-page memo set out why the government should refuse to grant Einstein a visa for a work trip to Pasadena:

> Albert Einstein believes in, advises, advocates or teaches a doctrine which, in a logical sense, as held by the courts in other cases, "would allow anarchy to stalk in unmolested" and result in "government only in name" . . .
>
> Albert Einstein believes in or is affiliated with Communist groups that advocate the overthrow by force or violence of the Government of the United States; he advocates "acts of rebellion" against the basic principle of all organized government . . .

he advocates "conflict with public authority"; admits that his "attitude is revolutionary"; that his purpose is "illegal" and that he intends to organize and lead, and collect money for and contribute money to a "militant opposition" and to "combat" the basic principle of our Constitution . . . he teaches and leads and organizes a movement for unlawful "individual resistance" and "acts of rebellion" against officers of the United States in time of war . . .

And who is the acknowledged world leader, who, by direct affiliation with Communist and anarcho-communist organizations and groups, and by his own utmost personal efforts, is doing most to "shatter" the "military machinery" for the defense of the existence of governments . . . ?

ALBERT EINSTEIN is that leader. Not even Stalin himself is affiliated with so many anarcho-communist international groups to promote this "preliminary condition" of world revolution and ultimate anarchy, as ALBERT EINSTEIN . . .

ALBERT EINSTEIN has promoted "lawless confusion" to "shatter" the Church as well as the State—and to leave, if possible, even the laws of nature and the principles of science in "confusion and disorder."

Rather than dismiss Mrs. Frothingham's letter, as they certainly could have done, the FBI took it as a call to action. The American consulate in Berlin was contacted, and Albert and Elsa were asked to report in person for some questions regarding their visa application.

"What is your political creed?" the interviewer began.

A little taken aback, Einstein stared at the man and then burst out laughing. "Well, I don't know. I can't answer that question."

"Are you a member of any organization?"

Albert raked a hand through his hair and looked at Elsa for help. "Oh yes! I am a War Resister."

"Who are they?"

"Well, they're my friends."

Forty-five minutes dragged by, with the process becoming no more smooth or pleasant. Eventually, Einstein lost his temper. "What's this, an inquisition?" he demanded. He reminded his interviewer that he hadn't chosen to go to America himself. "Your countrymen invited me. Yes, begged me. If I am to enter your country as a suspect, I don't want to go at all."

Having said his piece, he collected his hat and coat and left. Elsa immediately gave a report of the encounter to the papers, and let them know that if Einstein did not receive a visa by noon the next day he would cancel his trip to America. The consulate issued a statement saying a visa would be issued immediately.

68

Einstein and Elsa at the premiere of City Lights, *with its star,*
Charlie Chaplin, 1931.

On a twelve-acre estate next door to Charlie Chaplin in Beverly Hills, Mary Pickford and Douglas Fairbanks, at the time two of the most famous actors in the world, built a four-story, twenty-five-room house that they called Pickfair. It was surrounded by formal gardens and decorated with a decadence befitting old European aristocracy.

As far as American society was concerned, an invitation to Pickfair was as important as one to the White House. Among their wood-paneled rooms and eighteenth-century furniture, Fairbanks and Pickford would hold lavish, albeit sober, parties (Fairbanks was a teetotaler). Guests would admire the Chinese objets d'art and French paintings, the gilded niches and the Western saloon bar. Writers and politicians, actors, musicians, and heads of state would come to gossip, and compliment, and discuss spiritualism, the latest fad to capture Hollywood's attention.

Albert and Elsa were invited to a dinner party at the enormous mock-Tudor mansion. Dark, decorative curtains blocked out most of the California sunshine. Charlie Chaplin, who was on friendly terms with Einstein, was also there. Among the other guests was a prominent neurologist, who brought the conversation around to "thought transference," a form of telepathy of which he was quite convinced.

Einstein asked what it was.

"I think and concentrate my thinking on you," explained the brain specialist, "and you catch my thought."

"*Nein*, that's not possible."

"But wasn't your theory just as incredible—and still is to most people?"

Not at all, Albert insisted. Relativity was easy fare. To prove his point, he decided to demonstrate his theory with props. Slapping the edge of the dining table, he named it the edge of space. His plate played the part of either Earth or the sun—or maybe it was the entire universe, no one could really remember—and his knife and fork had something to do with the fourth dimension.

Try as she might, Pickford found that she couldn't follow a word Einstein said, partly because she was too in awe of her guest to ask for clarifications. After a while she stopped trying and, to amuse herself, concentrated on Douglas and Charlie, both of whom sat open-mouthed in complete, strained concentration and bewilderment. They had no luck following, either.

This lack of comprehension didn't stop the Hollywood royals from enjoying Einstein's company or inviting him to other parties, even if these evenings weren't always a great success. At one Chaplin soiree, Einstein met William Randolph Hearst and Marion Davies. Hearst, who was to become the inspiration for Orson Welles's *Citizen Kane*, owned magazines and newspapers in almost every major city in America. Davies, an actress with a talent for comedy, was Hearst's partner.

The party began well, but the conversation eventually slowed to a halt and Einstein sat looking off into the distance while Hearst stared down at his dessert plate. After a long silence, Davies, who had avoided talking to Albert for most of the evening, suddenly turned to him with a spark of malice. "Hallo!" she said, before waving her fingers over his head and asking, "Why don't you get your hair cut?" At this point, Chaplin declared it was time to move on to the coffee.

69

When, on January 30, 1933, Adolf Hitler took power as the chancellor of Germany, Einstein was in the middle of a term as a visiting professor at the California Institute of Technology in Pasadena. Einstein had initially underestimated the Nazis. Answering an American journalist's question about Hitler back in 1930, he had said, "He is living on the empty stomach of Germany." Once the economy improved, Albert guessed, he would no longer be of significance. But by the end of February 1933, it was clear that Einstein could not return to Germany, not least because his Berlin flat had twice been raided by the Nazis, once with his stepdaughter Margot inside.

The Reichstag had been set ablaze on the night of February 27, and the Nazis had used the incident to pass a law that suspended many constitutional protections—including the right to assembly, freedom of speech, and freedom of the press—and enabled political incarceration without specific charge and the confiscation of private property. Einstein mentioned to friends that he might have to move back to Switzerland. His decision to reject his home country was sealed by reports that his beloved summerhouse in Caputh had been ransacked and his boat confiscated, under the pretense of searching for a stash of Communist weaponry.

When he, Elsa, Helen Dukas, and his assistant Walther Mayer arrived in Belgium at the end of March after sailing from New York, Einstein went almost immediately to the German embassy in Brussels to hand in his passport and renounce his German citizenship. He retained his Swiss passport. When he stepped out of the embassy, he left German territory for the last time in his life.

At the time, however, it was not legally possible to surrender citizenship. Without the state's approval, one could hand in as many passports as one wanted and still be a German. Einstein's renunciation left the Nazis wondering what to do. At a meeting in August 1933, the Ministry of the Interior argued that, because of Einstein's fame, to publicly take away his citizenship would prejudice many against Germany. Better to quietly accept Einstein's action, which achieved the desired result. But a Gestapo officer argued that it was precisely because of Einstein's fame that the Nazis should expel him. He used his fame, it was argued, to spread lies and anti-German propaganda.

Albert had given the Nazis every reason to dislike him—beyond the obvious ones of being Jewish, internationalist, pacifist, and famous. He had made it clear that he did not support the Nazi Party in any way. During the run-up to the Reichstag elections of 1932 Einstein cowrote a manifesto warning that the country was in danger of becoming a fascist society. Elsa implored him not to sign any more political appeals. "If I were as you want to have me," he replied, "then I just wouldn't be Albert Einstein." And so he called for an anti-fascist alliance between the Social Democrat and Communist parties. His name headed posters bearing this message.

Einstein was officially declared to have forfeited his citizenship in 1934. The Nazis had wanted the pleasure of stripping Einstein of his right to be a German, but he had beaten them to it, and they were not pleased. In fact, Albert had twice bested the Nazis. For on the same day that he went to the embassy in Brussels in 1933, he had also posted a letter renouncing his membership in the Prussian Academy of Sciences—something else they had wanted to take away from him. "Dependence on the Prussian government," Einstein wrote in his letter, "is something that, under the present circumstances, I feel to be intolerable."

Max Planck was pleased with Einstein's decision, writing to him that it was the only way that would allow the relationship between Einstein and the academy to remain amicable. Planck had been worried that the academy would begin formal exclusion procedures against Einstein, an action that had been called for by some government ministers. "Even though on political matters a deep gulf divides me from him," he wrote to an academy secretary, "I am, on the other hand, absolutely certain that in the history of centuries to come, Einstein's name will be celebrated as one of the brightest stars that ever shone in the Academy."

The academy still went ahead with censuring their most famous member. The Nazis were furious that Einstein had anticipated them and expected something from the academy to satisfy their need for revenge. And so a statement was issued on the academy's behalf, accusing Einstein of "atrocity-mongering" and "activities as an agitator in foreign countries." There was no need, the statement ended, to regret the loss of such a member.

Only one academy member dared object to this treatment of Einstein: Max von Laue, Planck's old assistant, and a friend of Einstein since he had visited him in Bern in 1907. At a meeting on April 6, however, another good friend of Einstein's, Fritz Haber, went so far as to thank the secretary responsible for issuing the statement, and to label his action as appropriate.

By the time of this meeting, a law had been passed barring Jews from holding a state post, including at universities, where Jewish teachers and students had their academic identification cards confiscated. A month later, in front of the Berlin opera house, within view of the academy, nearly forty thousand people watched the burning of Jewish books in a huge bonfire. In the face of these acts, Einstein decided to divest himself of his memberships in all German organizations, asking Laue to act on his behalf.

Still desperate to humiliate Einstein, the Nazi authorities confiscated his summerhouse in Caputh, selling it to the local municipality. The initial plan was to have it serve as a camp for the Hitler Youth, but this failed to materialize due to lack of funds and in the end Einstein's beloved retreat was used by the government to train schoolteachers.

Although Einstein severed himself from his country, he did not sever himself entirely from his friends, even those who had not behaved as nobly as they ought to have. Planck made efforts to moderate the national antisemitic policies, at one point even writing to Hitler, to no gain, but for the most part he went along with the will of the government and encouraged other scientists to do the same, an action completely at odds with Einstein's beliefs.

"In spite of everything," Albert wrote to Planck in the midst of his dealings with the academy, "I am happy that you greet me in old friendship and that even the greatest stresses have failed to cloud our mutual relations. These continue," he went on, "in their ancient beauty and purity, regardless of what, in a manner of speaking, is happening further below."

70

Eduard and Albert Einstein in 1933,
the last time they saw each other.

In Belgium, Einstein rented a cottage in the seaside town of Le Coq sur Mer. Much was on his mind. He was due to visit Oxford in late May 1933, to give a lecture about the philosophy of science, but he wrote to his friend there, Frederick Lindemann, to ask whether he might delay the trip by a week. He felt compelled, he wrote, to visit his younger son, Eduard, in Switzerland, explaining that he couldn't stand the thought of not seeing him for six more weeks. He hoped Lindemann would understand.

Eduard was known as "Tete," or Teddy. He was very intelligent and quick-witted, but his had not been an easy life. He'd been sickly as a boy, and throughout his childhood he shuffled between doctors' visits and sanatoriums. He developed an interest in psychoanalysis, especially in the works of Sigmund Freud, and when he went to university it was as a student of medicine, in the hope of becoming a psychiatrist.

Signs of mental illness developed slowly, although they seem to have been exacerbated by a failed romance at university. By the time he was twenty, Tete had grown rambling, despairing, and angry. He once attempted to throw himself from a window, but was restrained by his mother. In the autumn of 1932, age twenty-two, he spent some time in an asylum near Zurich, being treated for schizophrenia.

Einstein was in Belgium when Eduard was readmitted, having shown no sign of recovery in the few months he had spent at home. "The sorrow is eating Albert up," Elsa wrote to a friend. "He finds it difficult to cope with, more difficult than he would care to admit. He has always aimed at being invulnerable to everything that concerned him personally. He really is so, much more than any other man I know. But this has hit him very hard."

When Einstein visited Eduard in the asylum, he brought his violin with him, and a photograph was taken of father and son. Tete loved music. Indeed, it was when he played the piano, which he did with an extreme, unsettling passion, that Eduard found something like clarity and calm. Einstein and his son had often played together, finding it easier to communicate with music than with words. There is no record of their conversation, but the visit did nothing to change Einstein's opinion that Tete's schizophrenia had been inherited from his mother's side and was something "about which nothing can be done." Though neither could know it, it was the last time Einstein would see his younger son.

After Mileva died in 1948, Einstein arranged and paid for Eduard to be cared for. All through his life, Albert felt the pressure of responsibility to ensure Eduard's welfare. At the same time, he didn't consider visiting him in Switzerland again, and as he grew older he was happy to hear less of him.

"You have probably already wondered about the fact that I do not exchange letters with Tete," he wrote to a friend. "It is based on an inhibition that I am not fully capable of analyzing. But it has to do with my belief that I would awaken painful feelings of various kinds if I entered into his vision in any way."

71

Commander Oliver Locker-Lampson was an angular sort of man: sharp mind, sharply dressed, with prominent cheekbones and an aquiline nose. Exceptionally charming when he wanted to be, but volatile in his moods, he was known as "one of those people who can get things *financed*." When he first met Einstein at Oxford in 1933 he was fifty-three and in the middle years of his long career as a Conservative MP, having first been elected in 1910.

He had been both a barrister and a journalist, and had accrued some sensational stories from his service in the First World War. He primarily served as commander of the Royal Naval Air Service's Armoured Car Division. In 1915, he and his squadrons were sent to Russia, to act in the service of the tsar. During his time on the Eastern Front, he befriended the Russian aristocracy, saved the life of a princess who had been shot in the neck, and was involved in a failed military coup against the provisional government in 1917. He later claimed to have been asked to murder Rasputin and assist the escape of Nicholas II after his abdication.

As might be imagined, he was no fan of socialism—for a while, indeed, he veered into stridently right-wing positions. After the war, he organized "Clear out the Reds" rallies and expressed admiration for foreign fascists. In a *Daily Mirror* article of September 1930, Locker-Lampson praised Hitler as a "legendary hero." In 1932, he sent Mussolini a pair of enameled cuff links and a record as a gift.

By 1933, however, his views had changed dramatically. Having been educated in Germany, he knew its language and people well and, after Hitler gained power, he became an early advocate of preparations for war against the Nazis. He redirected his energies to helping Jewish refugees find sanctuary, personally sponsoring many people, as well as working to help high-profile exiles such as Sigmund Freud and the emperor of Ethiopia, Haile Selassie.

Locker-Lampson liked to collect interesting and famous people. Although they were essentially strangers, Locker-Lampson wrote to Einstein after he'd returned to Belgium, asking if he would be his guest. Einstein was admirably at ease when it came to invitations, as happy to accept them as he was to issue them. In this instance, he had so enjoyed his trip to Britain (he had traveled to Glasgow after Oxford) that he decided to take Locker-Lampson up on his offer and returned to England in July 1933, only a month after his previous visit.

Keen to put on a good show for his star guest, Locker-Lampson arranged for him to meet some eminent, if out-of-office, politicians: the former foreign secretary Austen Chamberlain, the former prime minister David Lloyd George, and the former chancellor Winston Churchill. Locker-Lampson was particularly friendly with Churchill and took Einstein to Winston's family home, Chartwell, for a spot of Saturday lunch. Einstein wore a white linen suit, looking every bit the professor out for a summer jaunt, whereas Churchill looked more like a groundsman, sporting an oversize sun hat and coveralls. There is a photograph of the two men smiling in the gravel-pathed gardens.

During the visit, Albert raised the topic of providing sanctuary for German Jewish scientists at British universities, before warning

Churchill that Hitler was fixed on war and was already secretly preparing for it. Churchill replied, with Churchillian confidence, that Britain and America would certainly be able to check Germany's rearmament. Einstein was thoroughly convinced—one could almost say hoodwinked. As he wrote to Elsa later that day, "He is an eminently wise man." Characteristically, he added a note of mistaken political optimism: "It became clear to me that these people have taken precautions and will act resolutely and soon."

Einstein also attended a session of Parliament, watching from the public gallery as Locker-Lampson introduced a bill to extend citizenship to Jews fleeing from persecution. In his speech, Locker-Lampson recounted how, when signing the guest book at Lloyd George's house, Einstein had been forced to write that he had no address. "Germany," he went on, "has turned out her most glorious citizen—Einstein. It is impertinent for me to praise a man of that eminence. The most eminent men in the world admit that he is the most eminent. But there was something beyond mere eminence in the case of Professor Einstein. He was beyond any achievements in the realm of science. He stood out as the supreme example of the selfless intellectual. And today Einstein is without a home."

The bill never passed into law, but Locker-Lampson had certainly won over his guest.

72

In July 1933, while still in Le Coq sur Mer, Einstein received a cryptic letter that read "The husband of the second fiddler would like to talk to you on an urgent matter." It was a coded message regarding King Albert I of Belgium.

Einstein had got to know Queen Elisabeth of Belgium four years earlier, when he had visited Antwerp. They had spent the day playing Mozart together and discussing relativity over tea. He was invited back the following year, which was when he met the king. He and the queen played Mozart again and then he stayed for a private meal of spinach, eggs, and potatoes with the couple, in the absence of servants. They all liked each other immensely.

On receiving the letter, Einstein headed for the palace. The king had a request to make of the great scientist. Two conscientious objectors had recently been jailed in Brussels and a pacifist organization was making loud calls for Einstein to come to their defense. Einstein was well known as an outspoken pacifist. After the completion of his work on general relativity, his advocacy for peace had come to occupy as important a role in his life as science. For nearly twenty years, Einstein had been arguing that the only way to end war was for people, under all circumstances, to refuse military service. The king had called Albert to ask him not to speak out about these conscientious objectors.

Einstein agreed not to interfere. He did not want to displease the king—neither as a friend nor as the head of the country that was accommodating him in exile. But more significantly, in the face of the new political realities in Germany, he no longer believed war resistance was practicable.

"What I have to say will greatly surprise you," he wrote in a public letter addressed to the leader of the pacifist group. "Until quite recently we in Europe could assume that personal war resistance constituted an effective attack on militarism. Today we face an altogether different situation." He explained that Belgium was such a small country that it couldn't afford to misuse its armed forces, as "it needs them desperately to protect its very existence. Imagine Belgium occupied by present-day Germany!"

Einstein had an extraordinary confession to make. "Were I a Belgian," he wrote, "I would not, under the present circumstances, refuse military service; rather, I would enter such service cheerfully, in the belief that I would thereby be helping to save European civilization."

Although, he hastened to add that he had not abandoned pacifism: "This does not mean that I am surrendering the principle . . . I have no greater hope than that the time may not be far off when refusal of military service will once again be an effective method of serving the cause of human progress."

73

*Einstein with Commander Oliver Locker-Lampson, Marjory Howard,
and Herbert Eastoe standing guard, outside Locker-Lampson's cabin
in Roughton Heath, 1933.*

Einstein had a habit of lending his support to committees and societies without really checking particulars. Throughout his life he was a member, official supporter, or sometimes even the chairman of all manner of groups whose meetings he never attended. One of these committees had recently published a book—which he had not read—attacking Hitler's regime. As a consequence, the German papers, eager to attack the famous Jewish scientist, denounced "Einstein's infamy." More disturbingly, his name appeared on a list of Nazi assassination targets, and his picture was included along with those of other enemies of the German state, with the caption "Not yet hanged."

On August 30, 1933, Nazi extremists shot Theodor Lessing, the German Jewish philosopher and associate of Einstein, who had been staying in Czechoslovakia. The assassins were celebrated in Germany— Lessing's photograph had also been captioned "Not yet hanged." Within days, the press was reporting that Einstein was next on the list. The Belgian royal family assigned two police officers to stand guard at his rented house among the dunes. Einstein was irritated by the around-the-clock protection, often trying to give his guards the slip, but on the whole found everything rather amusing. On hearing a rumor of a $5,000 price on his head, he replied, touching his temple and smiling: "I didn't know it was worth that much!" He had "no doubt," as he told a Paris-based correspondent, that the threat was real, "but in any case I await the issue with serenity."

Elsa, by contrast, was anything but serene, and Albert did little to soothe her anxiety by arguing that "when a bandit is going to commit a crime he keeps it secret." They were leaving in a month or so for Princeton and the Institute for Advanced Study, and he felt things would be fine until then. It did not take long for Elsa to persuade her husband to go "on the run." She would stay behind and pack for America.

Einstein contacted Commander Oliver Locker-Lampson again. Only a month after his last visit, he took a boat back to Britain in the company of a journalist from the *Sunday Express*, who had been good enough to arrange the journey. Einstein apparently spent most of the passage working on his equations. He arrived in London on September 9 and from there was whisked away to Roughton Heath, near Cromer on the Norfolk coast, to stay in Locker-Lampson's cabin.

It was a tiny space, with walls made of long, thin logs, and it had a thatched roof. Locker-Lampson had brought along armed guards and two young women—introduced as his "assistants"—who, at least in a series of staged photographs, were also armed. "If any unauthorized person comes near, they will get a charge of buckshot," his host declared. Einstein assessed it differently: "The beauty of my guards would disarm a conspirator sooner than their shotguns."

For obvious reasons, the location of the great professor wasn't made public, but it would be hard to make the case that Einstein was isolated. In the course of his three-week stay, a number of guests came to visit him and he was interviewed by the *Daily Mail*. He even traveled to London to give a speech at the Albert Hall, in aid of displaced German intellectuals, with all nine thousand seats sold and people standing in the aisles.

74

For some reason, during Einstein's supposedly secret retreat, Locker-Lampson arranged for the pioneering modern sculptor Jacob Epstein to capture his guest's likeness.

The sittings took place in the cabin. It wasn't an ideal studio, not least because most of the space was taken up by a large piano, leaving little room to move. It was also dark. When Epstein asked the guards if they could remove the door to allow in more natural light, they sarcastically inquired if he wanted them to take off the roof. "I thought I should have liked that too," Epstein later recalled, "but I did not demand it as the attendant 'angels' seemed to resent a little my intrusion into the retreat of their Professor." One obstacle remained, however. "I worked for two hours every morning, and at the first sitting the Professor was so surrounded with tobacco smoke from his pipe that I saw nothing. At the second sitting, I asked him to smoke in the interval."

The shroud of tobacco done away with, the two men found they got on well. Einstein was full of bonhomie. He told jokes and, between sessions, would entertain Epstein by playing his violin or the intrusive piano. He told Epstein about the Nazis' efforts to discredit his work.

One hundred professors had apparently published a condemnation of relativity, prompting Albert to point out that if he were wrong, one professor would suffice just as well. He also reflected on how his sense of Jewishness had deepened because of Hitler's actions.

Epstein was a charismatic, if not handsome, man. His voice was sonorous and forceful, still with slight traces of his parents' Polish accent, and he dressed as the typical artist, in a beret and an oversize, almost rectangular dark suit. It did not take long before he won over the guards, and by his last day they all had beer together in the evening.

As a sitter Einstein was perfectly obliging. Past fifty, comfortable with his celebrity and ever indulgent, he would sit for almost anyone who asked to take his portrait and so was used to the process. Epstein was pleased with what he found.

Einstein appeared dressed very comfortably in a pullover with his wild hair floating in the wind. His glance contained a mixture of the humane, the humorous and the profound. This was a combination that delighted me. He resembled the ageing Rembrandt . . . Einstein watched my work with a kind of naive wonder and seemed to sense that I was doing something good of him.

And indeed, Epstein did do something good. His bronze bust is commonly regarded as one of his best portraits. It is rough, with thick lines and deep divots, the skin almost a series of ridges. It resembles the self-portraits of Rembrandt, except that while Rembrandt often looks as if he is tired of life, comfortable with defeat, Einstein is smiling at the corners of his mouth. Epstein managed to capture the intelligence of his subject, but he also understood that Einstein could not be separated from joy.

75

When, in October 1933, Einstein boarded the ocean liner *Westernland* in Southampton, ready to sail to America, he expected to return to Britain the following year, to spend another term at Christ Church. He never saw Europe again.

Rather, the Einsteins, along with Helen Dukas, quickly settled into their new American surroundings. Princeton may have been a "quaint and ceremonious village of puny demigods strutting on stiff legs," as Albert put it after a month of living there, but they still liked the place: its greenery, its buildings, its slightly European character.

"By ignoring certain social conventions," he said, "I have been able to create for myself an atmosphere conducive to study and free from distraction."

Since the turn of the century, Princeton University had been cultivating a reputation for academic excellence, and in mathematics in particular it had become a world leader. Einstein liked the freedom he was afforded at the institute to pursue his own research and saw it as part of a larger American insistence on the right to do as one wished, without the deference to tradition common in Europe. In April

1934, after only six months, Einstein announced he would be staying in Princeton indefinitely. He was made a full professor at the institute and his salary was upped to $16,000 a year.

For the most part, the Princetonians were respectful of their famous new resident, and left him alone as much as society dictated was polite. Though not intentionally, he seemed to lead his life now more as self-caricature. Anecdotes about him proliferated, most of which strayed into fiction while still capturing something of Einstein's essence.

He was known to help schoolchildren with their homework, and one Christmas Eve he borrowed a violin from a group of carolers and accompanied them on their rounds. He is also supposed to have made a habit of pulling a random book off the shelves of the institute's library, opening it at random, selecting a random paragraph, and then thinking about what he had read for three months, before returning to repeat the process.

In one story he is said to have tripped and fallen into a storm drain so deep that his head and arms protruded from the ground like a mushroom and he had to be helped out by a local photographer. Another tale involves somebody phoning the institute and asking to speak to one of the deans. The dean was not available, the caller was informed. In that case, came the tentative reply, would it be possible to have Dr. Einstein's home address? Unfortunately, that information could not be given out.

"Please don't tell anybody," whispered the voice on the phone, "but I *am* Dr. Einstein. I'm on my way home and I've forgotten where my house is."

76

In 1935, Einstein hit on a problem that he did not think quantum mechanics could account for. If two particles collided with each other briefly and then continued on their way, then by measuring, say, the momentum of one particle, you would be able to figure out the momentum of the other, even after the two had flown apart. What this meant was that it was possible to know something about a particle without measuring it. Within the laws of quantum mechanics this was not allowed. Quantum scientists therefore had to interpret the situation to mean that the act of measuring the first particle had some effect on the second particle, even if it was no longer nearby. To say this, Einstein argued, was absurd. In a letter to his friend Max Born, Einstein explained his position with a now famous phrase: "Physics," he wrote, "should represent a reality in time and space, free from spooky action at a distance."

Einstein's contemporaries got around this problem by saying that the two particles are "entangled"—that is, paired with each other. In certain ways, a pair of entangled particles act as if they are one system. The paper Albert coauthored on the subject is known as the EPR paper, after the initials of Einstein and his colleagues Boris Podolsky and Nathan Rosen. In this paper they were concerned with a particle's location and momentum, but the easiest way of explaining the effects of entanglement is to talk about a particle's "spin."

An electron's spin can exist in two states, either "up" or "down." One needn't worry about what this means exactly, it is important just that a particle can exist in one of either two states. In an entangled pair of particles, one will have "spin up" and the other "spin down." These properties are not built into the pairing. It is not that, as they go bumbling through the universe, one electron has the properties of an orange and the other an apple, but that they share the properties of both fluidly, in what is known as a "superposition." This only changes when one of the electrons is measured, when it interacts with something. At that point, the measured electron will be wholly and only an apple. Instantaneously, the other, unmeasured electron will be wholly and only an orange.

The reason this is important is the instantaneousness: the information that an entangled particle has suddenly fixed its spin is seemingly conveyed to its sister faster than the speed of light, no matter what distance separates them—even if they are at opposite ends of the observable universe. To Einstein this "non-locality" was not acceptable. He believed that nothing should be able to exceed the speed of light, not even information, otherwise the theory of relativity could not hold true. Luckily for Einstein, it was soon argued that, as the two particles should be regarded as the same physical entity, in fact no information is exchanged—there is no signal faster than lightning that is sent between them. As they are the same entity, a signal is not needed. Relativity survives, albeit only just.

Entanglement continues to be something of a thorn in the paw of physics, however, and no mouse has yet been able to take it out. Differing explanations of the phenomenon exist, but none has been settled on absolutely. There have also been many efforts to try to prove that, as seems logical, the spins of entangled electrons are fixed from the start, that one electron is always an orange and one an apple, and there is no fruity mixing going on. But they have not succeeded. Quite the contrary: entanglement is a proven phenomenon of nature. Quantum computers work because of it, and a photograph has been taken of entanglement at work.

Einstein's criticism was meant to show the bizarre consequences of following the logic of quantum mechanics, and to use this as evidence of the theory's falsehood. He did not believe that entanglement could truly exist, that by prodding one particle you could instantaneously affect another. But as it turns out, the world is far more absurd than Einstein thought.

77

Just as they had decided to stay in America, in May 1934 Einstein and Elsa received news that Ilse's health was failing. Elsa's daughter had been suffering from what was thought to be tuberculosis, but was in fact leukemia, and had moved in with her younger sister, Margot, in Paris.

Elsa set sail for Europe alone and arrived in Paris to find her emaciated daughter on the point of death. Ilse had for some time refused to receive proper medical care because she was convinced that her ailments were mainly psychosomatic, and she had instead submitted to a long course of psychotherapy. Elsa and Margot could do little for her except be with her as she died. Elsa was never quite the same. The experience devastated her spirit to the extent that she seemed to age by many years.

Margot moved to America to live with her mother and stepfather, leaving her husband behind, and in August 1935, Elsa and Albert bought a property in Princeton, across the road from their rented flat. Their simple white clapboard house at 112 Mercer Street was 120 years old. With four small square columns defending a diminutive porch, it was unostentatiously pretty, barely hidden behind a low garden hedge. They paid for it in cash and still had some money left over for renovations, which Elsa took charge of, even as her own health started to fail.

It became clear shortly after moving into their new home that Elsa would not be able to enjoy it for long. She developed a swelling of one of her eyes, something that tests in Manhattan revealed to be a symptom of heart and kidney problems. She was ordered to remain in bed, a treatment that worked in part, but she suspected she would never fully mend. When the summer of 1936 arrived, they rented a house next to Saranac Lake, three hundred miles north of New York, in the Adirondack Mountains. She felt sure that she would improve in such surroundings. "And if my Ilse walked into my room now, I would recover at once." The vacation did bring limited relief, but it was not a cure.

Although he read to Elsa occasionally, during her illness Einstein worked manically, sometimes hardly sleeping. Elsa informed her friend Antonina Vallentin that Albert was affected by the situation more than she had expected. "He wanders around like a lost soul," she wrote. "I never thought he loved me so much. And that comforts me." By the winter, Elsa was again bedridden. She died on December 20, 1936. And with her death, her husband cried. "Oh," he sighed, "I shall really miss her."

After a few days, he was back in the office, looking sallow and pale. Even so, one of his collaborators could not bear to offer him inane expressions of sympathy. Instead, they discussed a particular work problem, as if nothing had happened. Einstein produced two small but important papers in the month after his wife's death, but his early attempts to focus were sorry ones.

In a letter to Hans Albert, he wrote that he could not concentrate. Elsa's passing had made life difficult. "But as long as I am able to work," he went on, "I must not and will not complain, because work is the only thing that gives substance to life."

78

On one glowingly hot summer's day in 1937, Leopold Infeld and C. P. Snow drove to visit Albert Einstein at his rented vacation house on Long Island. Infeld was a Polish physicist who was collaborating with Einstein at Princeton on an equation to describe star movements. Snow was a molecular physicist at the University of Cambridge, as well as a novelist.

Snow described the encounter: "At close quarters, Einstein's head was as I had imagined it: magnificent, with a humanizing touch of the comic. Great furrowed forehead; aureole of white hair; enormous bulging chocolate eyes." He recalled that someone had once said it had "the brightness of a good artisan's countenance, that he looked like a reliable old-fashioned watchmaker in a small town who perhaps collected butterflies on a Sunday.

"What did surprise me was his physique. He had come in from sailing and was wearing nothing but a pair of shorts. It was a massive body, very heavily muscled: he was running to fat round the midriff and in the upper arms, rather like a footballer in middle age, but he was still an unusually strong man. He was cordial, simple, utterly unshy."

The group settled into conversation. Einstein asked Snow if he was a pacifist. "Far from it, I explained. I was by that time certain that war was inevitable. I was not so much apprehensive about war as about the chance that we might lose it. Einstein nodded."

The day was close, and very hot—it could be felt in the breath. There wasn't much to eat, though Einstein smoked his pipe incessantly. "Trays of open sandwiches—various kinds of wurst, cheese, cucumber—came in every now and then. It was all casual and Central European. We drank nothing but soda water. What with the heat and the sandwiches, I got as thirsty as if I had been dehydrated, and drank more soda water in eight hours than I normally did in eight months.

"Mostly we talked of politics, the moral and practical choices in front of us, and what could be saved from the storm to come, not only for Europe but for the human race. All the time he was speaking with a weight of moral experience which was different, not only in quantity but in kind, from anything I had met . . . It was something like talking to the second Isaiah."

Einstein spoke of the various countries he had lived in. As a rule, he preferred them in inverse proportion to their size. Did this mean he liked England? asked Snow. Yes, England—he liked England. It was a little like Holland, which he liked very much.

Why hadn't he chosen to live in England after his exile from Germany?

"No, no!"

"Why not?"

"It is your style of life." He laughed loudly. "It is a splendid style of life. But it is not for me."

Snow asked what he meant, and Einstein replied by saying that on his first day in England he had been taken to an estate in the country. There was a butler. There was evening dress. Following that, he had spent most of his time in England at Christ Church, which was positively overflowing with butlers and evening dress. The English, Einstein seemed to think, spent all day putting on and taking off different outfits. Snow objected, but Einstein wouldn't hear it. Had Snow heard of the German word *Zwang*? It meant, Albert explained, constraint, in the broadest possible way, constraint in any form, intellectual, emotional, societal. He wanted no *Zwang*.

They talked for many hours, until Snow saw the sky getting dark. "Einstein was talking about the conditions for a creative existence. He said that, in his experience, the best creative work is never done when one is unhappy. He could scarcely think of any physicist who had done fine work in such a state. Or any composer. Or any writer.

"It seemed a strange and unexpected remark."

79

Einstein had various pets in his household at 112 Mercer Street, all of which seemed to have their own problems with the world.

Bibo the parrot had many ailments. Einstein decided that poor Bibo was depressed and used to tell him jokes in an attempt to cheer him up.

And there was the fox terrier Chico, slightly chubby and quite disheveled, named after the Marx brother. "The dog is very smart. He feels sorry for me because I receive so much post. That's why he tries to bite the postman."

Einstein's tomcat, Tiger, was a sensitive fellow. He was miserable when it rained, and Einstein used to say to him, "I know what's wrong, my dear, but I really don't know how to turn it off."

Indeed, occasionally Einstein preferred animals to humans. When the cat of one of Einstein's friends, Ernst Straus, had kittens, Einstein was so eager to see them that he walked Straus back to his home and on the way became slightly worried—all of Straus's neighbors, it turned out, worked with Einstein at the Institute for Advanced Study. "Let's walk quickly," he said. "There are so many people here whose invitations I've declined. I hope they don't find out I came to visit your kittens."

80

Einstein wasn't in the habit of personally accepting honorary degrees. He received so many of them. However, in May 1946, he made an exception for Lincoln University in Pennsylvania, the first Black college to grant degrees in America. Trading one green and wooded campus for another, he made the short journey from Princeton so that he could teach a lesson about his relativity equations and also give a speech.

"My trip to this university was on behalf of a worthwhile cause," he announced to the students and faculty. "There is a separation of colored people from white people in the United States. This separation is not a disease of colored people. It is a disease of white people. I do not intend to be quiet about it."

And, true enough, he wasn't quiet about it. In essays and in speeches, Einstein openly challenged the racism he found embedded in America. He threw his support behind prominent Black intellectuals such as W. E. B. Du Bois and Paul Robeson, joined societies, and became the co-chair of the American Crusade to End Lynching. And it was not only in such public ways that he chose to stand against racial intolerance, but also with neighborliness and small private kindnesses.

On April 16, 1937, the famous contralto Marian Anderson arrived in Princeton to give a concert that evening. When she went to the Nassau Inn to book accommodations for the night, she was denied a room because she was Black. When Einstein heard of Anderson's situation, he invited her to stay with him. The two had met several years before when they had talked briefly backstage at Carnegie Hall after one of her performances. Einstein had loved her version of Schubert's "Death and the Maiden," while Anderson had admired his sensitive face and shock of white hair, and had felt humbled by him.

When Anderson arrived at Mercer Street, Einstein came downstairs to welcome her. Margot had prepared a room for her so that she could rest and change, and brought up a tray of food. Although Einstein did not go out as much as he used to, he attended the concert. She performed to a standing-room-only audience at Princeton's McCarter Theatre and showed "complete artistic mastery of a magnificent voice," according to the *Daily Princetonian*. "Miss Anderson had the audience at her feet from the first Handel aria to the last negro spiritual. It is hard to discuss such a performance without the excessive use of superlatives. Seldom is a voice like this combined with such a perfect intellectual and emotional understanding of the music."

After the concert there was a reception. As a matter of course, Einstein, the local celebrity, was invited, although as Anderson put it, no one "would have been offended had he begged off." But he came along for a while, and then waited up for Anderson to return home after him. From then on, whenever she sang in Princeton, she would stay with the Einsteins on Mercer Street.

On one occasion, Margot let Marian use her room for her to practice. Unknown to Anderson, the room was occupied by Bibo, the family parrot (who could say, "Beautiful, how beautiful" and "Give me a little kiss"). He was covered in his cage for the night, but as she began her exercises, as if out of nowhere, he began to sing along with her—"*Truu, truu, truu* . . ."—and Anderson burst out laughing.

81

The FBI file on Albert Einstein is 1,400 pages long. For more than twenty years the Bureau kept tabs on the professor, resorting on some occasions to opening mail, tapping phone lines, and even breaking into private property. As his file also includes anecdotes that agents had heard in conversation, with surprising frequency it takes on the tone of gossipy slander, as in this entry:

> EINSTEIN was alleged to be a personal courier from Communist Party Headquarters relaying messages orally to selected sources throughout the United States concerning important information being distributed by the Communist Party. These messages were of too great an importance to be trusted through the mails, telephone, telegraph or other means of communication and for this reason EINSTEIN being a trusted Communist was selected as the personal courier for the Party.

Another note suggests that Hans Albert was being kept as a hostage by the Soviets, and used as leverage to make Einstein an unwilling participant in Communist activities. This line of inquiry was dropped, however, when it was discovered that Hans Albert was not in Russia, as the Bureau had supposed, but working in Berkeley as a professor at the University of California.

Agents would consult an array of sources relating to the subject of their investigations. Einstein's file has surprisingly little of his political writing, but does include, for example, this report, based on a newspaper article once read by an agent:

> Einstein was one of many distinguished Germans who lent their influence and prestige to German Communists prior to the rise of Hitler . . . Einstein publicly declared, in 1947, that the only real party in France with a solid organization and a precise program was the Communist Party. In May, 1948, he and "10 former Nazi research brain trusters" held a secret meeting to observe a new beam of light secret weapon which could be operated from planes to destroy cities.

Action on the death ray was taken no further because, as the file notes, "The Intelligence Division of the Army subsequently advised the Bureau that this information could have no foundation in fact."

Sometimes members of the public would send letters to the FBI with information or tips, or often just suspicions. If a missive was deemed relevant, then it, too, would be included. Hence this informative postcard in Einstein's file, signed with the name "American":

> Are we safe with Atomic Energy as long as we have men like Einstein on our list? Watch out for him. [P.S. Flying saucers are SMALL experiments by Russia for disks 1000 times larger—later on.]

82

When the playwright Jerome Weidman was a young man, he was invited for an evening of chamber music at the New York home of a prominent philanthropist. He wasn't looking forward to it much, as he was nearly tone-deaf, and had always found it difficult to enjoy music. He sat down in the large drawing room, the band struck up, and he withdrew entirely into himself, trying to look as thoughtful as he could. When the band finished, he clapped along with everyone else.

To his right, a gentle voice asked, "You are fond of Bach?"

His neighbor was Albert Einstein.

Weidman was not fond of Bach. He knew nothing about Bach. He began to formulate a courteous response, but stumbled. It was only a polite question, and he could have easily answered in the same manner, inoffensively saying little. However, looking at Einstein's eyes, Weidman could tell that his answer would be taken seriously. He did not feel as if he could tell a lie.

"I don't know anything about Bach," he answered, embarrassed. "I've never heard any of his music."

Einstein was astonished.

"It isn't that I don't want to like Bach," Weidman assured him quickly. "It's just that I'm tone-deaf, or almost tone-deaf, and I've never really heard anybody's music."

Now Einstein was concerned. "Please, you will come with me?" he said.

He took Weidman's arm and led him out of the crowded room, drawing confused stares in their direction, and causing a ripple of hushed speculation. They went upstairs, to the study, where Einstein let go of his new acquaintance and shut the door.

"Now. You will tell me, please, how long you have felt this way about music?"

"All my life. I wish you would go back downstairs and listen, Dr. Einstein. The fact that I don't enjoy it doesn't matter."

"Tell me, please. Is there any kind of music that you do like?"

"Well, I like songs that have words, and the kind of music where I can follow the tune."

He smiled. "You can give me an example, perhaps?"

"Almost anything by Bing Crosby."

"Good!"

Einstein went to the phonograph and searched through his host's catalogue. "Ah!"

He put the record on: Bing Crosby's "Where the Blue of the Night Meets the Gold of the Day." Clearly pleased, Einstein smiled encouragingly at Weidman and kept tapping the air with the stem of his pipe in time with the music. After a little while, he stopped the record.

"Will you tell me, please, what you have just heard?"

Weidman attempted to sing back the lines as best he could. When he finished, Einstein was overjoyed.

"You see! You do have an ear!"

And so the two progressed with another record: "The Trumpeter" by John McCormack, and another, a refrain or two from a one-act opera.

"Excellent, excellent," said Einstein when Weidman managed to reproduce the string of notes he heard. "Now this!" And another record, and another, and another. Once Weidman had hummed some music without words, it seemed they were done. He couldn't believe that Einstein had paid him so much sincere attention.

"Now, young man," Einstein said. "We are ready for Bach!"

They returned to the drawing room shortly before the musicians began again.

"Just allow yourself to listen," he whispered reassuringly. "That is all."

And so Weidman listened to Bach's "Sheep May Safely Graze."

This time his applause was genuine.

After the musicians had taken their bows, Weidman and Einstein's hostess approached the two.

"I'm so sorry, Dr. Einstein," she said pointedly, "that you missed so much of the performance."

"I am sorry too," Einstein replied. "My young friend here and I, however, were engaged in the greatest activity of which man is capable."

"Really? And what is that?"

"Opening up yet another fragment of the frontier of beauty."

83

Einstein in 1939.

A DAY AT THE OFFICE, 1939

Morning at 112 Mercer Street: Maja, Margot, Helen, and Albert around the dining table, drinking the last of the coffee. He's wearing suit trousers and a loose white shirt, the collar of which is squished under a chunky ribbed sweater. His pipe is clenched in his mouth. His doctor has instructed him to give up smoking again, and so he restlessly grinds the stem between his teeth, in search of some trace of pleasure.

The breakfast done, they wish him well at work and, with no particular rush, he sets off across the small porch, down the three steps to the garden gate, and out onto Mercer Street. The trees that line the street have lately come into leaf. The gardens of his neighbors are vibrant. After five minutes or so, he is stopped by a woman on her way to town. She admonishes him for setting a bad example for the children of the neighborhood. Her daughter simply can't be persuaded to wear socks because she knows that he doesn't wear them. Einstein expresses his sympathy, but defends his position—and the intelligence of her daughter. Socks are entirely unnecessary, he assures her.

Past the McCarter Theatre and ten minutes farther through town takes him to Fine Hall, a redbrick, neo-Gothic pile belonging to the mathematics department of Princeton University. Only built in 1931, it nevertheless has oak paneling, leaded-glass windows, and a common room furnished with chess tables and leather armchairs. It serves as the temporary home of the Institute for Advanced Study, which will finish building its own campus in the autumn.

His two assistants are already in his office.

"Good morning, gentlemen," he says.

Bergmann and Bargmann—for such are their names—return the greeting.

"Good morning, Professor."

Einstein sits behind his desk and starts to rummage through a forest of paper. Not long ago, he was forced to abandon an attempt at a unified field theory he had been working on for six months, and Bergmann and Bargmann still seem dejected. He doesn't mind at all, though. They are working on a new approach and he is already confident he is on the right road. Indeed, he tells Bergmann and Bargmann that his new idea "is so simple God could not have passed it up." His smiles, good humor, and resolve quickly rub off on his assistants.

There is a mid-morning break for tea in the common room. Most of the others are here today. Clutching china cups and saucers, he and Hermann Weyl, Oskar Morgenstern, John von Neumann, and Eugene Wigner talk as colleagues, about their work, about nothing much.

Einstein does not see Niels Bohr, which he is glad about. Bohr has been visiting Princeton for a month or so, but Einstein has been avoiding him. They've exchanged some small talk, of course, but none of their usual back-and-forth regarding the quantum world. Nor have they discussed the news from Europe: that in Germany, before Bohr left for America, the physicists Otto Hahn and Fritz Strassmann had bombarded a uranium atom with neutrons and succeeded in splitting the atom. The faculty has since gone wild producing papers about this nuclear "fission," as it's being called. No, Einstein has said his piece in a recent lecture about his latest attempts toward a unified field theory, which Bohr attended. Looking directly at Niels, he'd said that he'd long tried to explain quantum effects via the method just described. Let it stand. Enough of the old arguments for now.

Back in his office, he is soon finished for the day, leaving his desk an even greater mess than he found it. No doubt somebody will eventually tell him off about it. He smiles at the thought. He walks home for lunch with three or four of his colleagues, sharing an intense discussion. They stop outside his house in the noon sun and continue to talk earnestly for some time, before each goes his own way and Einstein is left standing on the pavement alone. Lost in the aftermath of the conversation, he forgets himself and begins to walk back to the institute, until Miss Dukas rushes out of the house to call him in for his meal.

Lunch is a highly satisfying dish of macaroni, which brings to mind a time long past in Milan with his family. Dukas informs him of her day and hands him some of his mail, the letters she thinks it would be worth his addressing himself. There is also one from Eduard, she says, and she summarizes it for him. He can answer it another day, when he can face it.

He talks to his sister, who is a little fretful at the moment because she has recently discovered that she enjoys hot dogs, despite being a vegetarian. Albert, not wishing to see her worry, declares the sausage henceforth a vegetable. Maja is very pleased with this.

They talk of Germany, of course, of Hitler, and the war that will come. They worry about how Maja's husband, Paul Winteler, and Michele Besso and his wife, Anna, are faring in Italy under Mussolini. Einstein is thankful that Maja was able to get out of Florence and join him. And he is pleased that Hans Albert has moved to America. He's going to be thirty-five soon, remarks Maja. Really, Albert says, when is his birthday?

Einstein mourns the fact that he can no longer do as much as he used to for emigrants from Europe. The program he had set up to encourage Americans to help persecuted European Jews had come to nothing. And while for several years he had managed to sponsor individual immigrants to the US, with loans for travel expenses or monetary gifts, or with an affidavit attesting to their character, now his money has dried up.

He begins to withdraw from the conversation—his thoughts are elsewhere. After a nap, he stays in his study until well into the night, stopping for only a light supper of sandwiches. If he works through a few more things on this new field theory, he might just get there, and prove that he's right.

84

Einstein and David Rothman, Horseshoe Cove, Nassau Point, 1939.

For his summer vacation in 1939, Einstein stayed on the North Fork of Long Island. One day, while visiting the small town of Southold, he walked into the local hardware store and asked if he could have some sundials. The store's owner, David Rothman, apologized—he had to admit that they didn't sell them. But, he went on, he did have one in his garden. He led the professor to it and asked if it would do. Einstein laughed loud and deep, from his belly, before finally lifting up his foot: "Sundials." Because of his thick German accent, Rothman hadn't understood that Einstein was after a pair of sandals.

The two became friends. Rothman used to assemble local musicians to form string quartets for musical evenings at his house, and Einstein became a regular guest and contributor. On one such occasion, also present was a young Benjamin Britten. He accompanied the tenor Peter Pears on the piano, with Einstein on the violin. Both musicians remembered Albert's somewhat untrustworthy intonation.

Einstein was staying close by, on Nassau Point, a wedge of land jutting into Little Peconic Bay, composed of small beaches, sparse woods, and a sprawling set of gentle coves, all lined with reeds and trees and wooden docks. The house he rented from a local doctor was modest—two stories, a low roof, and a porch, with a view of the sea from a height.

He liked to walk in the surrounding woods, contemplating the unity of all forces and worrying about the advancing tide of Nazism, occasionally attracting the attention, but not the intrusion, of the locals. His time there wasn't lonely, however—he had his family. And the two-time Academy Award–winning actress Luise Rainer and her husband, the playwright Clifford Odets, were visitors. Einstein apparently flirted with Rainer so much that Odets cut his head out of some of their vacation photographs.

Above all, Einstein came to Nassau Point to sail. Alongside music, sailing was his purest pleasure, even though he couldn't swim and made a point of never wanting to learn. Children on Long Island often had to rescue him after he capsized his tiny sailboat, *Tinef*—which roughly translates from Western Yiddish as "junk." He once attempted to sail to Southold with his sister, Maja, so that Rothman could store his boat for him, but he overshot and missed the town completely. In the end, the siblings were out at sea for nine hours without any food before they were found.

Mostly, he let his boat drift in the idyllic waters of Peconic Bay, lost in his ideas and calculations. But when he was in the mood for fun, Einstein would fling himself toward other boats, only to pull away at the last minute, laughing at the startled local sailors.

85

About midafternoon, on Wednesday, October 11, 1939, Alexander Sachs was shown into the Oval Office for an audience with President Franklin D. Roosevelt.

"Alex," the president greeted him, "what are you up to?"

Sachs began, not for the first time in their acquaintance, with a parable. An inventor, he said, once visited Napoleon, claiming that he could build ships that needed no sail and that paid no heed to the wind. With such vessels Napoleon would be able to attack Britain regardless of the weather over the sea. "Bah!" was Napoleon's reply, before angrily sending the inventor away. Ships without sails, indeed. Here, though, was Sachs's punch line—the man who had addressed Napoleon was Robert Fulton, the inventor of the steamboat. What Sachs now had to say to Roosevelt, he continued, was every bit as important as Fulton's message.

Sachs reached for his papers. He was carrying an urgent letter from Albert Einstein.

Some months earlier, two Hungarian refugee physicists, Leo Szilard and Eugene Wigner, had taken a road trip to Long Island to track down Einstein on his vacation. They wanted his help. It was becoming increasingly certain that uranium could be used to help create exceptionally powerful bombs, and they were worried Germany was trying to buy considerable amounts of the element. Szilard and Wigner's best idea for stopping this was to ask Einstein to write a letter to his friend Queen Elisabeth of Belgium, because the largest deposits of uranium ore were to be found in the Congo, a Belgian colony. By writing to his "dear queen," they hoped that Einstein could dissuade the Belgian government from selling uranium to Germany.

After driving around and asking the locals if they knew where Dr. Einstein was staying, Szilard and Wigner eventually found him. Sitting on the small porch of his rented house, in the heat of mid-July, they explained their worries and talked him through the process of creating an explosive chain reaction in uranium.

Within fifteen minutes, Einstein had grasped the implications of such technology and agreed that he should write a letter, albeit to a Belgian minister he knew, rather than the royals. Wigner, showing some sense, pointed out that if three foreign refugees were going to write to another country's government about matters of defense, perhaps they should do so via the State Department. Einstein later wrote a draft of a letter in German, which Wigner translated and sent to Szilard. Through a mutual friend, Sachs got involved and offered to deliver the letter straight to the White House.

Now that they had a different recipient in mind, Einstein and Szilard revised their work. No longer was this a letter involving Congolese uranium deposits and Belgian exports, but rather one that called on the president of the United States to contemplate the practicalities of nuclear weapons:

I believe . . . that it is my duty to bring to your attention the following facts and recommendations:

In the course of the last four months it has been made probable—through the work of Joliot in France as well as Fermi and Szilard in America—that it may become possible to set up a nuclear chain reaction in a large mass of uranium, by which vast amounts of power and large quantities of new radium-like elements would be generated. Now it appears almost certain that this could be achieved in the immediate future.

This new phenomenon would also lead to the construction of bombs, and it is conceivable—though much less certain— that extremely powerful bombs of a new type may thus be constructed. A single bomb of this type, carried by boat and exploded in a port, might very well destroy the whole port together with some of the surrounding territory. However, such bombs might very well prove to be too heavy for transportation by air.

Einstein went on to point out where the world's best sources of uranium were and warn that the Nazis seemed to be collecting as much as they could. "I understand," he wrote, "that Germany has actually stopped the sale of uranium from the Czechoslovakian mines which she has taken over." Einstein suggested that the American government should procure uranium, set up links with scientists working on nuclear chain reactions, and speed up experimental work in that area.

"Alex," the president said over his brandy when Sachs had finished reading his summary of Einstein's words, "what you are after is to make sure the Nazis don't blow us up."

"Precisely."

Roosevelt called in his personal assistant, General Edwin Watson. "Pa," he said, for that's how the general was known, "this requires action."

86

Shortly before Einstein was formally stripped of his German citizenship by the Nazis, in April 1934, the United States Congress introduced a joint resolution to naturalize him. Their reasons for doing so, as set out in the resolution, were that Einstein was accepted as a "savant and genius," that he was an esteemed humanitarian, that he had publicly professed his love of the US and its Constitution, and that, above all, America was known throughout the world as a "haven of liberty and true civilization."

Einstein declined the offer. In fact, he was saddened and embarrassed by it. He only wished to be treated like any other new immigrant to the US, without honors and benefits. So when Albert decided to make Princeton his permanent home, he set about applying for American citizenship along the normal routes. As Einstein was still a Swiss citizen, it was not a lawful necessity for him to do this, but it was something he wanted to do.

The immigration visa Einstein needed could only be filed from a US embassy, the nearest of which happened to be in Bermuda. Therefore, in May 1935, he and his family sailed to the island for a few days, in what was to be his last trip outside the United States. On their arrival in Hamilton, the royal governor greeted them and recommended the island's two best hotels. Albert immediately disliked their grandiosity, and they ended up staying at a small guesthouse he spotted while wandering through the capital.

Einstein received a mass of invitations to official balls and receptions, all of which he declined, preferring to explore the island and go sailing with a German chef he had met at a restaurant. When Albert had not returned after seven hours out at sea, Elsa began to worry that the cook might be a Nazi sympathizer who had abducted her husband. But, hurrying to the chef's home, she found the two merrily enjoying a feast of German dishes.

While on Bermuda, Einstein didn't do a brilliant job at filling out his forms. In his Declaration of Intention he managed to get both the month and year of his and Elsa's wedding completely wrong. He made mistakes regarding where and when Elsa was born and was also wrong about Hans Albert's and Eduard's birthdays. His application was processed despite these errors, and five years later he took his citizenship test in Trenton, New Jersey.

As part of this process he agreed to be interviewed after his exam for the immigration service's radio program *I Am an American*. In the course of this, he argued that, to secure a future without wars, all nations, including America, would have to surrender part of their sovereignty to a global organization that would have complete control of all of its members' military power.

Along with Margot, Helen Dukas, and eighty-six others, he was sworn in on October 1, 1940. To the reporters covering the event, he praised his new country. America, he said, would prove that democracy is not just a system of government, but "a way of life tied to a great tradition, the tradition of moral strength."

87

84 Grange Loan
Edinburgh
July 15, 1944
Dear Einstein . . .

I had a kind of breakdown last winter from which I have not
quite recovered. It was the result of many causes: a little overwork,
the stress of the war in general and the extinction of the European
Jews, the transfer of my son to the Far East (he is after many
adventures quite safe on a pathological course in Poona, India),
etc. But the most depressing idea was always the feeling that our
science, which is such a beautiful thing in itself and could be such
a benefactor for human society, has been degraded to nothing but a
means of destruction and death. Most of the German scientists have
collaborated with the Nazis, even Heisenberg has (I learned from
reliable sources) worked full blast for these scoundrels—there are
a few exceptions . . . The British, American, Russian scientists are
fully mobilized and rightly so. I do not blame anybody. For under
the given circumstances nothing else can be done to save the rest of

our civilization . . . There must be a way of prohibiting a repetition of such things. We scientists should unite to assist the formation of a reasonable world order. If you have any definite plans, please let me know. I am rather powerless, sitting at this pleasant but backward place . . . Here in Britain it is very difficult to keep up connections with people. Traveling is possible only in the most urgent cases, and meetings in the South are restricted by the flying bombs.

But the military situation is excellent and we hope the European part of the war will soon be over . . .

I have tried, together with my Chinese pupil Peng, an excellent man, to improve the quantum theory of fields, and I think we are on the right track. Schroedinger, on the other hand, has improved your and other people's attempts to unite the different fields in a classical way. I think the next step should be a combination and merging of these two approaches. But I am too old and worn out to try it.

With kind regards and best wishes,

Yours ever,

Max Born

September 7, 1944

Dear Born,

I was so pleased about your letter that, to my surprise, I feel compelled to write to you, although no one is wagging a finger at me to do so . . .

Do you still remember the occasion some twenty-five years ago when we went together by tram to the Reichstag building, convinced that we could effectively help to turn the people there into honest democrats? How naive we were, for all our forty years. I have to laugh when I think of it . . .

I have to recall this now, to prevent me from repeating the tragic mistakes of those days. We really should not be surprised that scientists (the vast majority of them) are no exception to this rule, and if they are different it is not due to their reasoning powers but to their personal stature, as in the case of Laue. It was interesting to see the way he cut himself off, step by step, from the traditions of the herd, under the influence of a strong sense of justice. The medical men have achieved amazingly little with a code of ethics . . . The feeling for what ought and ought not to be grows and dies like a tree, and no fertilizer of any kind will do much good. What the individual can do is to give a fine example, and to have the courage to uphold ethical convictions sternly in a society of cynics. I have for a long time tried to conduct myself this way, with a varying degree of success.

Your "I feel too old . . ." I am not taking too seriously, because I know this feeling myself. Sometimes (with increasing frequency) it surges upwards and then subsides again. We can, after all, quietly leave it to nature gradually to reduce us to dust if she does not prefer a more rapid method . . .

We have become Antipodean in our scientific expectations. You believe in the God who plays dice, and I in complete law and order in a world which objectively exists . . . Even the great initial success of the quantum theory does not make me believe in the fundamental dice-game, although I am well aware that our younger colleagues interpret this as a consequence of senility . . .

With kind regards to you and your family (now freed from the flying bombs),

Yours,

A. Einstein

88

Sometime in the mid-1940s, Einstein and his assistant Ernst Straus had just finished preparing a paper. They were looking for a paper clip, to fasten their hard work together. They pulled open all manner of drawers and eventually found one, old and mangled beyond use. It must have been quite a heavy-duty thing, because they couldn't straighten it by hand. Einstein and Straus set about looking for a tool so that they could bend it into a usable shape. In their search, they came across a box of new, shiny, perfectly formed paper clips. Einstein took one and started to shape it into something with which to bend the old paper clip. Straus asked him what on earth he was doing, to which Einstein replied, "Once I am set on a goal, it becomes difficult to deflect me." Thinking for a moment, Einstein then said, "This would make a good anecdote about me."

89

In the spring of 1949, Niels Bohr visited the Institute for Advanced Study. When he arrived at the office of Abraham Pais, a friend and old colleague, he began by saying, "You are so wise . . ." Pais laughed, understanding immediately what he wanted. Bohr famously struggled with formulating sentences, at least in writing, and he often found it helpful to have someone to help him work through how to express his ideas. Bohr was having difficulty with a paper he had agreed to write to celebrate Einstein's seventieth birthday. It was not the first time he had asked Pais to play the role of sounding board.

The two went downstairs to Bohr's office, which was not actually Bohr's office at all—he was borrowing it from Einstein for the duration of his visit. Einstein found the room too big and preferred to work in the smaller adjoining office, which belonged to his assistant. Striding through the door, Bohr told Pais, "Now, you sit down. I always need an origin for my coordinate system." Pais obliged, sitting at a large table while Bohr paced furiously around, stooping slightly as he tried to force out some sentences. Pais would jot down any that were eventually arrived at. It was not a quick process. Bohr would sometimes spend several minutes stuck on one word, having to drag the next part of what he wanted to say out of himself.

Naturally enough, as Bohr was writing an article about his friend, at one point he came to rest on the word "Einstein." Pacing so quickly that he was close to running around the table, he began repeating, "Einstein . . . Einstein . . . Einstein . . ." After a while, he moved to the window, where he looked out at Princeton, still muttering, "Einstein . . . Einstein . . ."

Pais noticed the door opening very softly. Einstein tiptoed into the office and, putting his finger to his lips, motioned Pais to be silent. Smiling naughtily, still on tiptoe, Einstein made his way toward the table. Meanwhile, Bohr was still standing by the window, punctuating the silence: "Einstein . . . Einstein . . . Einstein . . ." Bohr suddenly turned back into the room, with one last, forceful "Einstein," only to find himself face-to-face with the man himself, as if he had conjured him there by magic.

Bohr, who seldom lost his composure, blushed to his roots. Einstein explained that he had been heading for Bohr's tobacco pot, which sat on the large table. His doctor, he told Bohr, had forbidden him from smoking. As usual, he had chosen to take this to mean that while he could no longer buy any tobacco, stealing it was fine. The three of them burst into laughter.

90

During a speech in Wheeling, West Virginia, in February 1950, Senator Joseph McCarthy waved a sheet of paper at his audience. Written on it, he declared, were the names of 205 workers within the Department of State who were members of the Communist Party. So far as can be established, this was a lie.

Over the next few weeks, the number of subversives claimed to be at work vacillated considerably. At one point it was fifty-seven, then eighty-one, and at another time only ten. Truthfulness hardly mattered, however. McCarthy's speech caught the attention of the American people and chimed with their fears. After all, America's Cold War foreign policy had suffered some recent shocks: the Communist Party had won the Chinese Civil War the previous year, for example, and the Soviets had successfully detonated a nuclear device.

The Republican Party saw McCarthy's narrative of a Communist plot inside government as a vehicle to regain political power. In 1952, the Republicans were granted a sweeping victory, winning control of the House and the Senate as well as the presidency. McCarthy was nominated the chair of the Permanent Subcommittee on Investigations, responsible for rooting out suspected Communists, not just in government but across large sections of society.

McCarthy had a particular interest in purging the educational system of those he regarded as enemies of his idea of America. As part of this effort, in April 1953 the subcommittee called before it William Frauenglass, a teacher from a high school in Brooklyn. He was considered a disloyal American because of a course for other teachers he'd taught six years earlier called "Techniques of Intercultural Teaching," which sought to explore ways in which teachers might ease intercultural or interracial tensions in the classroom. A witness claimed the course and its teachings were "against the interests of the United States." When Frauenglass was asked which organizations he belonged to, he refused to answer.

Frauenglass wrote to Einstein to ask for a statement educators might rally behind. In a public letter, Einstein advised Frauenglass to refuse to testify. "This kind of inquisition violates the spirit of the Constitution. If enough people are ready to take this grave step, they will be successful. If not, then the intellectuals of this country deserve nothing better than the slavery which is intended for them." Einstein assured Frauenglass that he would gladly go to prison himself to protect the freedoms that the government, in its paranoia, sought to dismantle.

McCarthy was not pleased. He said that any American who advised citizens to keep secrets from their government, whoever they were, was a bad American, a disloyal American, an "enemy of America." Many people took McCarthy's message to heart and the Institute for Advanced Study received letters criticizing Einstein as un-American and suggesting he move to Russia. Unperturbed, Einstein continued to speak out. Democracy would not last long, he warned, if such attacks on the freedoms of teaching and opinion continued.

The spirit of inquisition that had overtaken the American government reminded him of the Germany he had left behind in the early

1930s. Indeed, so disgusted was Einstein with the political climate toward teachers and scientists that, in 1954, in a letter to a magazine, he only half-jokingly wrote, "I would rather choose to be a plumber or a peddler in the hope to find that modest degree of independence still available under present circumstances."

Inevitably, plumbers from around America responded to Einstein. He was offered membership in the Chicago plumbers union and received work tools in the mail. An enterprising New York plumber named Stanley Murray wrote to him with a proposal:

> *Since my ambition has always been to be a scholar and yours seems to be a plumber, I suggest that as a team we would be tremendously successful. We can then be possessed of both knowledge and independence.*
>
> *I am ready to change the name of my firm to read: Einstein and Stanley Plumbing Co.*

91

Kurt Gödel and Albert Einstein in Princeton, 1954.

Kurt Gödel is often cited as the greatest logician of all time, and his greatest achievements are his two incompleteness theorems. In short, together these proved that within any mathematical system there will always exist certain propositions about numbers that cannot be proved when adhering to the rules of the system. To simplify it further, one could say that Gödel managed to prove that not all of mathematics could be proved. While this surprising result didn't really affect the everyday work of mathematicians, it had implications for the philosophy of mathematics as a whole, and, along with relativity, it helped contribute to the intellectual feeling, so prevalent in the first half of the twentieth century, that the foundation stones of the reliable and the known were far less secure than supposed.

Talking to Gödel was widely viewed as a terrifying experience. It seemed that whenever one of his colleagues at the Institute for Advanced Study approached him with a subject of conversation—seemingly any subject—they would find he had already thought about it, and thought about it at length, to the point where he had seen to its end and could preempt everything his colleague had to say about it. He always dressed well and was said to have no sense of humor, though his favorite film was *Snow White and the Seven Dwarfs*. He suffered from hypochondria and paranoia, to such an extent that he refused to leave the house when prominent mathematicians were in town, for fear that they might try to kill him. He and his wife moved house several times, as he believed that bad air was emanating from the appliances, and he also believed in ghosts. In his later life, he would eat very few foods, and then only after his wife had tried them. Mostly, he subsisted on a diet of baby food and self-prescribed laxatives and antibiotics.

And yet, for all this, Gödel's good friend at the institute was Albert Einstein. The two would walk to and from work together, discussing ideas. Einstein once said jokingly that the only reason he bothered going to work at all was so that he could walk with Gödel. Unlike so many of Albert's colleagues, Gödel had no difficulty in arguing with him. One of the things they talked about was time. Gödel was so interested in time that he published a paper on general relativity in which he proposed a solution to Einstein's field equations that produced a rotating universe. He managed to show that in such a universe time travel would be possible in a manner consistent with relativity. He didn't do this to lend credence to the idea of time travel, however. His argument was that if the absurdity of time travel could exist, even in a hypothetical universe, then time itself could *not* exist.

Toward the end of 1947, Gödel was due to attend his American citizenship hearing. Characteristically, he had prepared exceptionally well for the test—too well, in fact. For months leading up to the hearing, he taught himself about the history of the settlement of North America, and through that about the culture and history of various Native American tribes. He studied the town of Princeton in detail, learning who the mayor was, how the township council operated, how the borough council was elected, and so on. He also committed himself to fully understanding the US Constitution, and was excited, if horrified, to find a logical flaw in it. He had, he believed, discovered a way in which one could, legally, abiding by the Constitution, establish a fascist dictatorship in the United States. He thought this would be an excellent thing to bring up during his exam.

Einstein and another friend of Gödel, Oskar Morgenstern, were to be witnesses at the hearing. They did not think it an excellent thing to mention the possibility of an American dictatorship during a citizenship hearing. Morgenstern drove Gödel to the test, swinging by Mercer Street to collect Einstein on the way. Gödel sat in the back and Einstein got in the front. As they drove through the wintry trees and gray fields of Princeton, Einstein shifted in his seat to look back at his anxious friend.

"Now, Gödel," he said with a sardonic smile, fully aware of the extraordinary effort Kurt had put into his studies, "are you *really* well prepared for this examination?"

Just as Einstein thought it would, this question immediately set Gödel into a worried panic that perhaps he hadn't prepared well enough after all. After calming him down, Morgenstern and Einstein then spent most of the car trip trying to dissuade Gödel from discussing his perceived flaw in the Constitution.

The judge for the hearing was, thankfully, on friendly terms with Einstein.

"Now, Mr Gödel," he asked after the three professors had settled themselves before him, "where do you come from?"

"Austria."

"And what kind of government did you have in Austria?"

"It was a republic, but the constitution was such that it was finally changed into a dictatorship."

Einstein and Morgenstern began to worry.

"Oh! This is very bad," the judge replied. "This couldn't happen in this country."

They braced themselves for disaster.

"Oh, yes, I can *prove* it."

Gödel launched into an explanation, but the judge, responding to the look on Einstein's face, quieted poor Kurt and let him know that he really needn't go into all that. He was finally made an American citizen, and took his oath, on April 2, 1948.

No one ever thought to make a note of the flaw he had spotted in the Constitution.

92

Chaim Weizmann—the Zionist leader who had been responsible for Einstein's tour of America in 1921 and had gone on to become the first president of Israel—died in November 1952. The Jerusalem evening newspaper *Maariv* recommended as Weizmann's successor "the greatest Jew alive: Einstein."

It was a powerful suggestion, and at the time it sounded more than fine to David Ben-Gurion, the prime minister of Israel, who publicly endorsed the idea as quickly as he could.

He sent an urgent telegram to Israel's US ambassador, Abba Eban. Eban wired Einstein, inquiring if he would allow someone from the embassy to visit him in Princeton to convey an important message.

Einstein was aware of what this meant. The American newspapers had reported Weizmann's death and had recommended Einstein as his successor. At first he had thought it was a joke. Einstein did not want the position. As he said to Margot, "If I were to be president, sometimes I would have to say to the Israeli people things they would not like to hear."

Seeing little point in some poor official having to drive all the way to Princeton, he rang Eban expressly to ask the ambassador not to offer him the job.

"I am not the person for that and I can't possibly do it."

"But I can't tell my government that you phoned me and said no," Eban said. "I have to go through the motions and present the offer officially."

Eventually Einstein relented, realizing that it would be insulting to refuse the invitation before he had even received it. Someone from the embassy was soon dispatched.

"Acceptance," the formal letter informed him, "would entail moving to Israel and taking its citizenship. The Prime Minister assures me that in such circumstances complete facility and freedom to pursue your great scientific work would be afforded by a government and people who are fully conscious of the supreme significance of your labors."

Eban was anxious to express that the offer embodied "the deepest respect which the Jewish people can repose in any of its sons . . . I hope that you will think generously of those who have asked it, and will commend the high purposes and motives which prompted them to think of you at this solemn hour in our people's history."

Einstein's reply, which he handed to the minister as soon as he arrived, read:

> *I am deeply moved by the offer from our State of Israel, and at once saddened and ashamed that I cannot accept it. All my life I have dealt with objective matters, hence I lack both the natural aptitude and the experience to deal properly with people and to exercise official functions. For these reasons alone I should be unsuited to fulfill the duties of that high office, even if advancing age was not making increasing inroads on my strength. I am the more distressed over these circumstances because my relationship to the*

Jewish people has become my strongest human bond, ever since I became fully aware of our precarious situation among the nations of the world.

In the end, Ben-Gurion was more than thankful for the refusal. While waiting for Einstein's answer, he had begun to harbor doubts.

"Tell me what to do if he says yes!" he jokingly asked his assistant. "If he accepts we're in for trouble."

Had Einstein accepted, the State of Israel would have had a president who was ill-disposed to authority, formality, and paperwork, who was outspoken, who did not speak Hebrew, had not had a bar mitzvah ceremony, whose views on God were notoriously unorthodox, and who had been a vocal opponent of the creation of a Jewish state. "I would much rather see reasonable agreement with the Arabs on the basis of living together in peace than the creation of a Jewish state," he had once said, in succinct summary of his view of Zionism. "Apart from practical consideration, my awareness of the essential nature of Judaism resists the idea of a Jewish state with borders, an army and a measure of temporal power, no matter how modest. I am afraid of the inner damage Judaism will sustain." Trouble indeed.

Einstein met Eban at a black-tie reception in New York two days after the business had been concluded. As Eban noted, Einstein was not wearing socks.

93

Albert's closest friend, Michele Angelo Besso, whom he had known for over fifty years, died on March 15, 1955, the day after Einstein's seventy-sixth birthday. Besso's son and sister wrote to Einstein to give him the news. He sent them a letter in return, thanking them and reflecting on who his friend had been. He died less than a month later.

Princeton, March 21, 1955
Dear Vero and dear Mrs. Bice,

It was truly very kind of you to give me, in these difficult days, so many details about Michele's death. His end was in harmony with his whole life, and with the circle of his loved ones. This gift of a harmonious life is seldom paired with such a sharp intelligence, especially to the degree in which it was found in him. But what I most admired in Michele, as a man, was the fact that he managed to live for many years not only in peace but in lasting consonance with a wife—an undertaking at which I twice rather shamefully failed.

Our friendship began when I was a student in Zurich; we met regularly at music events. He, the elder and a scientist, was there to stimulate us. The circle of his interests seemed simply boundless. However, it was critico-philosophical concerns that seemed to win him over.

Later, we were reunited by the patent office. Our conversations on our way home were of an incomparable charm—it was as if the contingencies of daily life simply didn't exist. In contrast, we later had more difficulty understanding each other in writing. His pen could not keep up with his versatile spirit, such that it was, in most cases, impossible for his correspondent to guess what he had failed to write down.

Now he has again preceded me a little in parting from this strange world. This has no importance. For people like us who believe in physics, the separation between past, present and future has only the importance of an admittedly tenacious illusion.

I am sending you my sincere thanks and my best thoughts.

Yours,

A. Einstein

94

Einstein, a few days before his seventy-fifth birthday, 1954.

In 1948, Einstein was diagnosed with an aneurysm of the abdominal aorta. He was told that this would likely be the death of him. "The strange thing about growing old is that the intimate identification with the here and now is slowly lost," he wrote to one of his friends. "One feels transposed into infinity, more or less alone."

On the afternoon of April 13, 1955, Einstein collapsed. The day before, his assistant, noticing him grimace, had asked if everything was all right. Yes, he had replied, everything except himself. Helen Dukas called the doctor and he was given morphine so he could sleep. More doctors came the following day. The aneurysm had started to break, but Einstein refused surgery. "It is tasteless to prolong life artificially," he explained to Dukas. "I have done my share. It is time to go. I will do it elegantly."

He was taken to the hospital the next day, after Dukas had found him in bed, in agony, unable to lift his head. His condition improved to such an extent that he asked for paper, pencils, and his glasses so that he could do some work in his hospital bed. He talked with Hans Albert, who had flown from San Francisco to see him, about physics, and his friend Otto Nathan about politics. He looked over a draft of a speech he was to give for Israel's Independence Day, and he wrote twelve pages of equations, complete with crossings-out and amendments, still hoping to find his unified field theory.

His recovery, however, was fleeting. Shortly after one in the morning on Monday, April 18, the night nurse on duty, Alberta Roszel, noticed a difference in Einstein's breathing and heard him muttering very quietly. The aneurysm had burst and death would claim him soon. But Roszel did not speak German, and so his last words were lost.

The funeral took place the same day as his death. There were twelve guests, including Hans Albert, Helen Dukas, Otto Nathan, and Einstein's girlfriend, Johanna Fantova. Few of them wore black. In the sharp brightness and chill of spring, under incongruous sunlight, Nathan read from Goethe's memorial poem for the playwright Friedrich Schiller. Goethe describes his friend's almost unearthly talents, his courage, his "unchanging ever-youthful glow"; and he mentions his dedication to fighting against the injustices of society. It was an appropriate choice of text. Einstein was—to friends and strangers, at least—the most generous and gentle of people, but that did not detract from the hardness in him. He was convicted, near immovable, in his belief that the evils of society should be spoken out against and fought to the best of one's ability.

"He gleams like some departing meteor bright," Nathan ended. "Combining, with his own, eternal light."

And that was effectively all the ceremony Einstein was given, much as he had wished. He had wanted as little public veneration as could be achieved. He had been careful not to leave behind places tied to the idea of him. His office at the institute was to be used by others; the house on Mercer Street was to be sold and lived in. And he made it clear that he wanted no marker of his body, no piece of ground for the mighty man to lie beneath. His ashes were scattered in an undisclosed place.

95

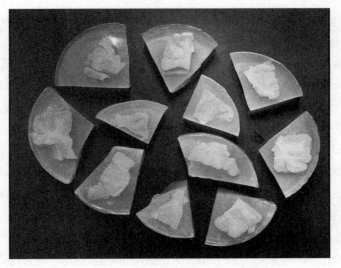

*Einstein's brain, dissected, segmented, and preserved in celloidin,
ca.1980.*

A t the time of Albert Einstein's death, Thomas Harvey was a
pathologist at Princeton Hospital. He was a Quaker. With his
short haircut and high hairline, he looked convincingly ordinary. More
unkindly, one might call him forgettable. It was Harvey's job to perform the routine autopsy on Einstein's body. While an understandably
upset Otto Nathan looked on, Harvey removed and examined each of
Einstein's major organs. He then replaced them before sewing up the
body—or rather, he replaced almost all of them. In the autopsy room,
Harvey decided, entirely without permission, to keep Einstein's brain.

When this was discovered a few days later, Einstein's friends and family were outraged. Hans Albert tried to complain, but Harvey argued that Einstein would have wanted to serve a scientific use. Hans Albert, unsure what he could practically do about the situation, begrudgingly accepted the state of affairs. With this retroactive approval in place, Harvey was soon approached by the US Army's pathology unit, but he refused their various requests to meet with them, instead choosing to cut up the brain, embalm it, and store it in some glass cookie jars.

He left his job at Princeton, taking the brain to the University of Pennsylvania, where it was carved into 240 pieces and preserved in celloidin, a hard, rubbery substance. After driving the pieces home in the back of his Ford, he then put them away, floating in their jars, in his basement. He divorced his wife, remarried twice, and moved about the country, always taking the brain with him and often leaving no forwarding address. In Wichita, Kansas, he worked as a medical supervisor in a biological testing lab and kept the brain in a box meant for apple juice cartons, near a beer cooler, buried under old newspapers.

In Weston, Missouri, he practiced medicine and attempted to study the brain, but in 1998 he lost his medical license after failing a competency exam. In Lawrence, Kansas, he worked in a plastic-extrusion factory, on the assembly line, and moved into an apartment next door to a gas station. Here he befriended his neighbor, who turned out to be the beat novelist and poet William S. Burroughs. The two would meet regularly for drinks on Burroughs's porch and swap stories. Burroughs proudly told his other friends that he could have a piece of Einstein's brain any time he asked.

All the while, Harvey had been periodically sending off portions of the brain—some slivers mounted on slides, some larger chunks— to a smattering of researchers across the country. His choices of who received these cerebral gifts were pretty random, mostly based on whoever's work appealed to him at the time, although sometimes he did answer requests for samples. He sent a mayonnaise jar full of various bits to a neuroscientist at the University of California, Berkeley, for instance. But Harvey rarely asked that the recipients *do* anything with his present.

Einstein's brain has not ended up as a scientific marvel, but rather as something akin to a religious relic. It is the tongue of St. Anthony, the heart of St. Camillus, preserved so that we may observe and venerate it as a tangible reminder of someone more than human. Contrary to Hans Albert's wishes, it has become an object of popularism and tourism. There's an app of the brain available, should one want it, consisting of a "brain atlas" constructed from slides and photographs. And some of Harvey's slices have found their way to the Mütter Museum in Philadelphia, where they nestle, very comfortably, in the company of a malignant tumor removed from President Grover Cleveland's mouth and a piece of tissue from the neck of John Wilkes Booth.

96

At Europe's Spaceport, in the jungle outside Kourou, French Guiana, on June 5, 2013, the European Space Agency's Automated Transfer Vehicle-4 was blasted skyward. The unmanned cargo freighter was named *Albert Einstein*. About the size of a double-decker bus, with four solar array wings sticking out of a tubular body, its purpose was to resupply the International Space Station.

When it docked, ten days after liftoff, *Albert Einstein* delivered food, water, oxygen, and propellant to the astronauts of the ISS. It also carried a 3D-printed toolbox, gas masks, and a new water pump and recycling unit, as well as equipment for science experiments. All in all, the delivery weighed seven tons and amounted to more than 1,400 items, including a space food treat—tiramisu.

Among this bounty was a copy of the first page of the manuscript for general relativity. This was signed aboard the station, some 250 miles from the surface of Earth, by the astronaut Luca Parmitano, in a symbolic gesture of respect and debt. Because, of course, without the equations of general relativity to refer to, space exploration would be markedly more difficult. It is essential, for example, to take into account the effects of relativity when determining the correct orbits of heavenly bodies or spacecraft as they explore the solar system, or keeping track of interplanetary probes via the radio signals they send out.

Indeed, where accuracy in space is concerned, Einstein's theory has to be allowed for. We see the practical upshot of this most obviously in the Global Positioning System. The satnav in your car or on your phone receives signals from a system of satellites orbiting Earth around twelve thousand miles away, each broadcasting its location and the exact time. Your car then uses the differences in the time it takes to receive those signals to work out its respective distance from each of the satellites. This then reveals its own location on Earth. Time is crucial to the whole operation. But because the satellites are so far away from the planet, they experience a weaker gravitational effect, which is to say that, by a tiny fraction, the satellites experience time moving slightly faster than on Earth. If GPS didn't have general relativity built into it, to adjust for the satellites' clocks being always slightly faster than their counterparts on Earth, your phone could easily send you miles in the wrong direction.

97

The first test of a hydrogen bomb took place in 1952, at Enewetak Atoll in the Pacific Ocean, and the operation was code-named Ivy Mike. The bomb itself was given a nickname: "the sausage." The explosion contained the power of more than 10 million tons of TNT. The fireball created was approximately two miles wide, and within seconds a mushroom cloud had spread out to a diameter of one hundred miles, covering the blue sea. The blast caused waves up to twenty feet high. The nearby islands were stripped of vegetation, and radioactive coral debris fell on ships thirty-five miles away.

At the time, Edward Teller, who had been perhaps the greatest supporter of the hydrogen bomb, was nearly five thousand miles away, in Berkeley, California, home to most of America's atomic research, but he watched the shock of the explosion register on a seismometer. He quickly telegrammed a colleague in Los Alamos, the message reading only "It's a boy."

Drones carrying filter paper were flown through the radioactive clouds. The material these collected, as well as tons of coral from the atoll, were all sent to Berkeley. Analysis confirmed that in the intensity of the explosion, a new element had been created. Among the mass of debris studied in the lab, around a hundred atoms of element 99 were detected.

The element is silvery, soft, and metallic, and glows blue in the dark. One gram contains one thousand watts of energy. Like all the actinides—those exotic elements right at the end of the periodic table—it is heavy, and very reactive. Its various isotopes have half-lives ranging from mere seconds to more than a year, so even in the best of circumstances it is short-lived. It also holds the distinction of having no practical use at all.

Ivy Mike was, of course, a close secret, and the results weren't declassified for three years. In the *Physical Review* of August 1, 1955— three and a half months after Einstein's death—the discovery of element 99 was finally published. In the article, the discoverer, Albert Ghiorso, and his colleagues proposed that the new element be named after Einstein. At the time it was rare to name an element after a scientist, although not unprecedented: in 1944, curium (atomic number 96) had been named after Einstein's old friend Marie Curie and her husband, Pierre.

Einstein, for all his discomfort over the subject, was one of the fathers of quantum mechanics. Moreover, his paper on Brownian motion in 1905 had been one of the first to treat the existence of atoms as a reality—in fact, his paper essentially mathematically *proved* the existence of atoms. These two things alone would have justified his inclusion in the table of elements. Of course, as element 99 had been born in the rush and rumble of atomic destruction, there was also a relevance in naming it after the man responsible for $E=mc^2$, the equation that made sense of the destruction of the hydrogen bomb, and the man who had written to President Roosevelt encouraging him to invest in nuclear chain reactions. It also served as a memorial, a way of honoring the death of the most famous scientist in the world.

Even so, the choice couldn't help but be ironic, given Einstein's re-gret over the danger of nuclear weapons and his limited practical role in their creation. Because of the FBI's suspicions of Albert, he wasn't given clearance to know anything about the American government's nuclear program, before or after the Second World War. He never worked on the bomb, and the Manhattan Project began much more as a result of Washington hearing about calculations made in England that showed an airborne nuclear bomb was feasible than because of the letter Einstein had written. He once declared outright that he was not "the father of the release of atomic energy," as he was sometimes perceived.

Much of his later life was spent advocating the establishment of a world government to safeguard peace for the future, and when, in 1950, he heard of President Truman's decision to develop the hydro-gen bomb, he made his first appearance on the new medium of televi-sion. Filming a statement in Princeton that was broadcast to the nation the next day on *Today with Mrs. Roosevelt*, he said that if the efforts to produce the H-bomb succeeded, "annihilation of all life on earth will have been brought within the range of what is technically possible. The weird aspect of this development lies in its apparently inexorable character. Each step appears as the inevitable consequence of the one that went before. And at the end, looming ever clearer, lies general annihilation."

The last public act of Einstein's life was to sign a manifesto calling on leaders to renounce war in an age of nuclear weaponry.

The symbol for element 99 is Es. Einsteinium.

98

Since 1929 we have known that the universe is expanding. Einstein was overjoyed at this discovery, as it allowed him to do away with a mathematical term he'd introduced into his equations of general relativity in order to have them describe a static universe—the cosmological constant (written as the Greek letter Λ, lambda). Before Einstein introduced this mathematical fix, relativity had in fact seemed to insist that the universe was expanding.

After the expansion of the universe was known about, it seemed logical to think of that expansion as slowing down. This assumption was based on the fact that if there is mass in the universe, then there is also gravity, and gravity is attractive—it "pulls" at things. Every galaxy inevitably pulls at every other galaxy. The expansion of space, then, was reasoned to be decelerating for effectively the same reason that the speed of an apple thrown up into the air decelerates. Both would fight against the pull of gravity.

This idea seemed so obvious and plausible that it was believed for nearly seventy years. But in 1998 it was shown, to considerable surprise, to be wrong. Two research teams, one led by Saul Perlmutter and the other by Brian Schmidt and Adam Riess, studied distant supernovas out in the universe. These majestic, exploded stars are known to have a standard brightness, so the fainter a supernova appears in the sky, the farther away it is. Once a distance is established, the age of the galaxies that house these cosmological beacons can be calculated—the farther away a galaxy is, the farther back in time it is. Using this knowledge, Perlmutter, Schmidt, and Riess were able to track the expansion of the universe through much of its history. They deduced that it had been expanding at a slower rate in the past than in the present. In other words, they discovered that the expansion of the universe is accelerating. For their work, they were awarded a joint Nobel Prize in 2011.

Gravity, of course, didn't suddenly disappear with this discovery. Each galaxy still pulls at every other galaxy. The reason the expansion of the universe is accelerating, therefore, has to be because something is counteracting the effects of gravity. Just as a rocket ship needs thrusters to propel itself upward, so the universe requires some kind of energy to act against the pull of galaxies.

It was realized much later that the term Einstein had introduced into his equations to try to keep the universe static—the cosmological constant—was ironically exactly the right tool to use to calculate the recently measured acceleration of the universe, specifically to calculate the "something extra" that counteracts gravity. It's worth explaining that the cosmological constant is not a constant like pi, for example. Like pi, it will not change with time, but unlike pi, it is not something humans can determine a priori, from only theoretical knowledge. At least at present, it must be measured. Measurements over the last twenty years have provided strong evidence that the accelerated expansion is driven by a cosmological constant and have determined, with only small uncertainties, its value. What Einstein had done when he thought he no longer needed it was to set the term to zero.

We might ask what the cosmological constant represents. Gravity can be thought of in physical terms, and if lambda is a kind of anti-gravity, then it must also have some physical significance, rather than being something purely mathematical. What it corresponds to in the real world is called "dark energy." Many scientists would argue that dark energy is the energy present in the vacuum of space. We think of a vacuum as exactly that—an emptiness through which only radiation passes. In the ocean of space, away from the merry dance of stars, there are blacknesses in which there appears to be nothing at all. But even here, there is still energy, fluctuating as if it were composed of millions upon millions of microscopic waves. Dark energy is thought to be the total energy inherent in the vacuum of space.

This theory leads to an estimate of the value of the cosmological constant as 1×10^{113} joules of energy per cubic meter. This is a huge number. However, based on observations of the expansion rate of the universe, it appears that the amount of dark energy in a cubic meter should be closer to 1.5×10^{-9} joules—which is a minute number. The discrepancy between the theoretical value of dark energy and observational evidence is a major concern in contemporary physics, part of what's called the cosmological constant problem. What it effectively signifies is that dark energy is barely understood at all.

The problem of dark energy is one of the challenges that the theory of relativity faces. Black holes present another—at the center of every black hole there lies a singularity, a point of such intense curving of space-time that the equations of relativity break down. Black holes have been photographed and gravitational waves emitted from them have been detected. They form an important part of the makeup of the universe, and yet they cannot be fully understood with Einstein's theory.

Indeed, while relativity for the moment still stands as the foremost way to make sense of the cosmos, it seems unlikely that it will stand as such forever. The state of contemporary physics bears some resemblance to the state of physics in Einstein's youth, when physicists would take turns in attacking and defending the work of Isaac Newton, some trying to patch the holes in his theory of gravity, while others (like Einstein) tried to tear it down. Now it is Einstein who has his detractors and defenders.

The universe outwits us. It outwitted Einstein during his lifetime, both with quantum mechanics—which undermined his notion of what the world should be like—and with his attempts at creating a unified field theory. In the final sentences of his final autobiographical reflection, in March 1955, only a month before his death, Einstein admitted that perhaps the past twenty years of his work had been in vain after all. "It seems doubtful altogether," he wrote, "whether a field theory can properly account for the atomistic structure of matter and radiation as well as of quantum phenomena." But he comforted himself that "the search for truth is more precious than its possession."

Outlining his personal beliefs, Einstein once wrote, "The fairest thing we can experience is the mysterious. It is the fundamental emotion that stands at the cradle of true art and true science. He who knows it not and can no longer wonder, no longer feel amazement, is as good as dead, a snuffed-out candle."

Motivated to puzzle out the mysteries of his age, Einstein was forced to reconceive the very nature of light, of time and space, of the character of the cosmos itself. In doing so, he revealed a more faithful picture of reality, but not one free of enigmas. And it is now by tackling the problems of relativity that a deeper understanding of the universe may be reached. We are blessed to live in a time in which the mysteries facing us are those of Einstein's universe.

99

SELF-PORTRAIT, AT THE AGE OF FIFTY-SIX

Of what is significant in one's own existence one is hardly aware, and it certainly should not bother the other fellow. What does a fish know about the water in which he swims all his life?

The bitter and the sweet come from the outside, the hard from within, from one's own efforts. For the most part I do the thing which my own nature drives me to do. It is embarrassing to earn so much respect and love for it. Arrows of hate have been shot at me too, but they never hit me, because somehow they belonged to another world, with which I have no connection whatsoever.

I live in that solitude which is painful in youth, but delicious in the years of maturity.

SOURCES AND ACKNOWLEDGMENTS

In researching a life of Albert Einstein, there is no better place to start than with the writings of the man himself. To this end, the project of *The Collected Papers of Albert Einstein*, published by Princeton University Press, is entirely invaluable. As an indication of the reliability and thoroughness of the *Papers*, the first volume was published thirty-six years ago, and with their most recent—sixteenth—volume, they have so far covered Einstein's writings to May 1929.

No complete collection of Einstein's scientific writings so far exists, although a complete list of his scientific publications can be found in either Albrecht Fölsing's biography or *Albert Einstein: Philosopher-Scientist*, edited by Paul Arthur Schilpp. Einstein's nonscientific writings, which are models of concise, eloquent expression, have been collected in *The World as I See It* (later expanded as *Ideas and Opinions*) and *Out of My Later Years*, as well as the voluminous *Einstein on Peace*, edited by Heinz Norden and Otto Nathan. This last contains almost all of Einstein's thoughts on pacifism. Also, Princeton University Press has recently collected together all of Einstein's attempts at autobiography (of which there are a few short examples) in *Einstein on Einstein*, edited by Hanoch Gutfreund and Jürgen Renn.

Einstein was a prolific, thoughtful, and highly entertaining letter writer. Some books of his correspondence with individual acquaintances have been published and all are valuable and good reading. Of these I used *The Born–Einstein Letters: The Correspondence Between Albert Einstein and Max and Hedwig Born from 1916 to 1955; Letters to Solovine, 1906–1955;* and *Einstein: Correspondance avec Michele Besso, 1903–1955.* Einstein's correspondence with Besso, originally published in German with a parallel French translation, has, for some reason, still not been published in English.

Apart from this, there is an abundance of Einstein biographies. The ones I relied on most heavily included *Einstein: The Life and Times* by Ronald W. Clark; *Einstein: His Life and Times* by Philipp Frank; *Albert Einstein: Creator and Rebel* by Banesh Hoffman and Helen Dukas; *Einstein: Profile of the Man* by Peter Michelmore; *Subtle Is the Lord: The Science and the Life of Albert Einstein* by Abraham Pais; *Albert Einstein: A Biographical Portrait* by Anton Reiser (the pseudonym of Einstein's son-in-law Rudolf Kayser); Carl Seelig's *Albert Einstein: A Documentary Biography; The Drama of Albert Einstein* by Antonina Vallentin; and *Einstein: The Man and His Achievement,* edited by G. J. Whitrow.

The list in the previous paragraph is of books published forty years ago or more. They all contain some notable inaccuracies, mostly not through any fault of their own, but because the facts were unavailable to the authors. Of the more recent biographies, *Einstein's Greatest Mistake: The Life of a Flawed Genius* by David Bodanis provides a brisk walk through Einstein's life and science. Then there is *Einstein: A Biography* by Jürgen Neffe; *Albert Einstein: Chief Engineer of the Universe,* edited by Jürgen Renn; and *Einstein: A Life* by Denis Brian.

Otherwise, *Albert Einstein: A Biography* by Albrecht Fölsing and *Einstein: His Life and Universe* by Walter Isaacson were vital references. Between these two books, there is hardly an episode in Einstein's life that is not covered. If the reader is searching for a comprehensive evaluation of Einstein, they can do no better than to start with Fölsing and Isaacson.

There is an exception to their completeness, however. This book contains an incident of Einstein's life that has not been written about before in the popular literature on Einstein, because it was not yet known about: the rekindling of his relationship with Marie Winteler, the letters relating to which were published in volume 15 of *The Collected Papers* in 2018. Also published were many of his teenage love letters to Marie, which made clear that his feelings for her had been deeper than previously thought.

A comprehensive bibliography of Einstein studies is beyond the intended scope of this book. I am, however, confident in saying that for any aspect of Einstein's life one could possibly wish to know more about, there exists a book in print. For example, if one wishes to learn more about his private life, then one could turn to *The Private Lives of Albert Einstein* by Roger Highfield and Paul Carter, which looks at Einstein's life almost purely in terms of his personal relationships. There is the brilliant *Einstein in Love: A Scientific Romance* by Dennis Overbye, which covers Einstein's early life and relationship with his first wife; or the two biographies of Mileva Marić: *Im Schatten Albert Einsteins* by Desanka Trbuhović-Gjurić and *In Albert's Shadow* by Milan Popović. *Einstein's Daughter: The Search for Lieserl* by Michele Zackheim details what is known about his forgotten daughter.

If geographical precision is more appealing, then there is *Albert Einstein in Bern* by Max Flückiger; *Einstein in Bohemia* by Michael D. Gordin, about his time in Prague; and *Einstein in Berlin* by Thomas Levenson. Or there's *Einstein on the Run: How Britain Saved the World's Greatest Scientist* by Andrew Robinson and *Einstein in America: The Scientist's Conscience in the Age of Hitler and Hiroshima* by Jamie Sayen.

The famed scientist's views on religion are accounted for in *Einstein and Religion: Physics and Theology* by Max Jammer. The First World

War is dealt with in *Einstein's War: How Relativity Triumphed Amid the Vicious Nationalism of World War I* by Matthew Stanley. More on the FBI can be found in *The Einstein File: J. Edgar Hoover's Secret War Against the World's Most Famous Scientist* by Fred Jerome. It was through Jerome's freedom-of-information requests that Einstein's FBI file became available to the public.

There are many explanations of Einstein's science to choose from, not least his own book *Relativity: The Special and the General Theory*. Beyond that, Russell Stannard's *Relativity: A Very Short Introduction* is a good starting point. *Einstein 1905: The Standard of Greatness* by John S. Rigden is a marvelously detailed and yet understandable explanation of the scientific papers from Einstein's annus mirabilis; and more specific still is $E = mc^2$: *A Biography of the World's Most Famous Equation* by David Bodanis. For reflections on Einstein by his colleagues, see *Encounters with Einstein* by Werner Heisenberg, *Writings on Physics and Philosophy* by Wolfgang Pauli, or *Einstein: A Centenary Volume*, edited by A. P. French.

And so the list goes on. It is worth remarking that of the innumerable books that take Einstein as their subject I haven't yet come across one that is not intelligent and useful.

Part of the joy of writing this book was the opportunity to look in detail at minor—you could say trivial—aspects of Einstein's life that would be inappropriate inclusions if featured in a more traditional, comprehensive exploration. The story of Einstein and Marian Anderson comes from her *My Lord, What a Morning: An Autobiography*; the Jacob Epstein anecdote is from his *Let There Be Sculpture*; the incident with William Golding is recalled in his essay "Thinking as a Hobby"; Bertrand Russell's entertaining opinions feature in an interview with John Chandos at Russell's home in North Wales in 1961; the tale of C. P. Snow visiting Einstein on vacation comes from the essay "On Albert Einstein"; and Jerome Weidman, in the essay "The Night I Met

Einstein," tells of Einstein's kindness as a music teacher. Einstein's time among the stars of Hollywood is found in Charlie Chaplin's *My Autobiography*, as well as Mary Pickford's *Sunshine and Shadow*. The information about the development of the electric lamp in the first chapter was drawn from multiple sources, including *Nature*, *IEEE Industrial Electronics Magazine*, the *Encyclopaedia Britannica*, and the Thomas A. Edison Papers; more about the spacecraft *Albert Einstein* can be found on the websites of the European Space Agency, NASA, and collectSPACE, and in *The Road to Relativity: The History and Meaning of Einstein's "The Foundation of General Relativity"* by Hanoch Gutfreund and Jürgen Renn; the macabre story of Einstein's brain is told in detail by Carolyn Abraham in *Possessing Genius: The Bizarre Odyssey of Einstein's Brain*.

"A walk to the office, 1925" I drew from an essay written by Esther Salaman in the *Listener* magazine in 1955. As with the "day at the office" chapters, this is lightly fictionalized. I have gathered up material from the year and employed false narrative cohesion to create a representative day for that time in Albert's life. For instance, I have Einstein saying that an idea is "so simple God could not have passed it up" to his assistants Bergmann and Bargmann, when he is instead recorded as saying this to another of his assistants, Ernst Straus. Nevertheless, I still aim to be factual—the talk about the rhinoceros mentioned in "A day at the office, 1904," for example, was found in *News from the Society of Natural Sciences in Bern from the Year 1904*; it took place on October 22.

Occasionally the research for these minor episodes led me to pleasant places. Einstein's friendship with David Rothman and the anecdotes about his time on the North Fork of Long Island can be found in *My Father and Albert Einstein* by Joan Rothman Brill; in *Letters from a Life, Vol. 2, 1939–45: Selected Letters and Diaries of Benjamin Britten*, edited by Donald Mitchell and Philip Reed; and in information held in the Southold Historical Museum. In Southold, next to what was Rothman's Department Store, there is now Einstein Square, where small, rough-hewn benches surround a plinth supporting a white marble bust of an aged Einstein with heavy jowls. Both a mural and a plaque proclaim Einstein's vacation there as his "happiest summer ever."

Should one be interested, it is worth taking the trip to Potsdam to see the Einstein Tower, not least because it sits in the utterly intriguing Albert Einstein Science Park, full of an abundance of working scientific buildings, dispersed throughout a wooded campus. Here brick Gothic observatories dot the landscape like wizards' castles, and one can see unexplained, rusted sheds that somehow help to calculate longitude and latitude, as well as twenty-one different types of industrial thermometers enclosed in a picket fence as if they might run off.

In my research I also made use of several articles from journals and magazines, and many newspaper articles. YouTube also provided some fascinating video footage and interviews—which revealed to me among other things that Albert possessed a much higher voice than I imagined for him. The Albert Einstein Archives at the Hebrew University of Jerusalem were exceptionally useful, as were the State Archives of Zurich, the Secret State Archives Prussian Cultural Heritage Foundation, the Niels Bohr Library and Archives at the American Institute of Physics, and the archives at ETH Zurich, the Institute for Advanced Study, and the Max Planck Society.

There is a host of people without whom this book would not have been possible, in particular: Michael Dine, who very kindly looked over the scientific sections of this book, clarified the physics, and explained some of the more technical issues of cosmology; my agent, Toby Mundy; and my editors, Georgina Laycock, Caroline Westmore and Katharine Morris, Rick Horgan, Dan Cuddy, and Olivia Bernhard, and their teams at John Murray and Scribner. Thank you for your astute edits and comments, which helped shape and thoroughly improve my various drafts. Thank you also to Juliet Brightmore, for her exceptional picture research and reassurances. My sincere thanks to the Albert Einstein Archives at the Hebrew University of Jerusalem, especially to Chaya Becker, and to Lisa M. Black at Princeton University Press. Also, to my colleagues at the *Times Literary Supplement*, particularly Robert Potts, who have been wonderfully accommodating and kind as I undertook this project. And, of course, thank you to my wife, Isabelle, without whose forbearance in the face of explanatory and biographical scraps, and whose understanding, questioning, and patience, I certainly should not have managed.

CREDITS

PERMISSIONS

All usages from *The Collected Papers of Albert Einstein* and the Albert Einstein Archives. © 1987–2021 by Hebrew University of Jerusalem. Published by Princeton University Press and reprinted by permission.

Quotations from Alice Calaprice, ed., *Dear Professor Einstein* reproduced with permission of Rowman and Littlefield Publishing Group Inc. through PLS-clear.

Quotation from *Baltimore Afro-American* courtesy of the AFRO American Newspaper Archives.

Quotations from Carl Seelig, *Albert Einstein: A Documentary Biography*, and Carl Seelig, *Albert Einstein: Leben und Werk eines Genies unserer Zeit*, permission from the Robert Walser Centre.

Quotation from *Commentary*, courtesy of Stephanie Roberts/*Commentary*

Quotation from Jacob Epstein, *Let There Be Sculpture: An Autobiography* © The estate of Sir Jacob Epstein/Tate.

Quotation from Max Born's letter of July 15, 1944, courtesy of Sebastian Born/The Max Born Literary Estate.

Quotation from Subrahmanyan Chandrasekhar, "Verifying the Theory of Relativity," *Bulletin of Atomic Sciences* reprinted by permission of Taylor & Francis Ltd, and *Bulletin of Atomic Sciences*.

Quotation from *The Listener*, courtesy Ralph Montague/Immediate Media Company.

Quotation from "The Night I Met Einstein" by Jerome Weidman originally

IMAGES

217 The Shelby White and Leon Levy Archives Center, Institute for Advanced
    Study, Princeton, NJ
225 Alamy Stock Photo/Sueddeutsche Zeitung Photo
249 Library of Congress/Corbis/VCG via Getty Images
253 Courtesy of the Leo Baeck Institute, New York. Image Ref: F 15280
269 Photo by Leonard McCombe/The LIFE Picture Collection/Shutterstock
278 Photo by Steve Pyke/Getty Images
281 Getty Images/Bettmann

QUOTATIONS

References to *The Collected Papers of Albert Einstein*, vols. 1–16 (Princeton University Press, 1987–2021) are indicated below with the abbreviation "*CPAE*." For each of these, I give the item information, followed by the volume number, item number (where applicable), and the page number of the relevant English translation supplement.

References to items in the Albert Einstein Archives at the Hebrew University of Jerusalem are indicated by "AEA," followed by their object number.

INTRODUCTION

viii   *"I knew the theory was correct"*: Ilse Rosenthal-Schneider, *Reality and Scientific Truth: Discussions with Einstein, von Laue, and Planck* (Wayne State University Press, 1980), p. 74.

ix   *"How excellent the postal"*: Antonina Vallentin, *The Drama of Albert Einstein* (Doubleday, 1954), p. 11.

ix   *"radiance of his large eyes"*: Einstein, *Letters to Solovine: 1906–1955* (Philosophical Library, 1987), p. 7.

xiii   *"Wait a minute, I've nearly finished"*: Carl Seelig, *Albert Einstein: A Documentary Biography* (Staples Press, 1956), p. 104.

xiv   *"strenuous intellectual work"*: Einstein to Pauline Winteler, May (?), 1897, *CPAE* 1:34, p. 32.

xiv   *"own obituary"*: Einstein, "Autobiographical Notes," in *Einstein on Einstein: Autobiographical and Scientific Reflections*, ed. Hanoch Gutfreund and Jürgen Renn (Princeton University Press, 2020), p. 157.

PARTICLE 2

5 *"Yes, but where are its wheels?"*: Maja Winteler-Einstein, "Albert Einstein: A Biographical Sketch," *CPAE* 1, p. xviii.

6 *"the obvious arbiter"*: Ibid.

6 *"the dopey one"*: Ernst Straus, "Reminiscences," in *Albert Einstein, Historical and Cultural Perspectives: The Centennial Symposium in Jerusalem*, ed. Gerald Holton and Yehuda Elkana (Princeton University Press, 1982), p. 419.

6 *"Father Bore"*: Erik H. Erikson, "Psychoanalytic Reflections on Einstein's Centenary," in ibid., p. 172.

7 *"threw a large bowling ball"*: *CPAE* 1, p. xviii.

7 *"it takes a sound skull"*: Ibid.

PARTICLE 3

9 *"something deeply hidden"*: Einstein, "Autobiographical Notes," in *Einstein on Einstein*, p. 159.

9 *"Young as I was"*: Gerald Holton, "On Trying to Understand Scientific Genius," *American Scholar* 41:1 (1971–2), pp. 95–110.

PARTICLE 4

10 *"ancient superstition"*: Ronald W. Clark, *Einstein: The Life and Times* (Harper, 1972), p. 25.

10 *"You never know"*: Anton Reiser, *Albert Einstein: A Biographical Portrait* (Albert & Charles Boni, 1930), p. 28.

11 *"nothingness of the hopes"*: Einstein, "Autobiographical Notes," in *Einstein on Einstein*, p. 157.

11 *"religious paradise"*: Ibid.

12 *"popular scientific books"*: Ibid.

PARTICLE 5

13 *"one of those poor people"*: Frank, *Einstein*, p. 8.

14 *"your mere presence here"*: Seelig, *A Documentary Biography*, p. 15.

15 *"a so-called 'child prodigy'*: Albin Herzog to Gustav Maier, September 25, 1895, *CPAE* 1:7, p. 7.

15 *"My failure seemed completely justified"*: Einstein, "Autobiographical Sketch," in *Einstein on Einstein*, p. 144.

16 *"a simple seriousness"*: Einstein, "Autobiographical Sketch," AEA 29-212.1.

16 *"left unforgettable impressions"*: Ibid.

PARTICLE 6

17 *"it is beautiful"*: Einstein to Marie Winteler, February 18, 1896, *CPAE* 1:16g (in vol. 15, p. 7).

17 *"the doughnut project"*: Einstein to Winteler, February 3, 1896, *CPAE* 1:16b (in vol. 15, p. 3).

18 *"Guess what? Today I played"*: Ibid., p. 4.

18 *"Without having read this letter"*: Postscript to Einstein to Winteler, April 21, 1896, *CPAE* 1:18, p. 13.

18 *"Your little basket arrived"*: Winteler to Einstein, November 4–25, 1896, *CPAE* 1:29, p. 29.

18 *"My love, I do not quite understand"*: Ibid.

19 *"the stupid little teapot"*: Winteler to Einstein, November 30, 1896, *CPAE* 1:30, p. 30.

19 *"I beseech you"*: Einstein to Winteler, before May 21, 1897, *CPAE* 1:33a (in vol. 15, p. 14).

19 *"I cannot come to visit you"*: Einstein to Pauline Winteler, May (?), 1897, *CPAE* 1:34, pp. 32–3.

PARTICLE 7

21 *"The words or the language"*: Jacques S. Hadamard, *An Essay on the Psychology of Invention in the Mathematical Field* (Dover, 1945), p. 142.

22 *"One would arrive at a time-independent"*: Einstein, "Autobiographical Sketch," in *Einstein on Einstein*, p. 144.

22 *"the first childlike thought experiment"*: Ibid.

PARTICLE 8

26 *"I owe more to Maxwell"*: Esther Salaman, "A Talk with Einstein," *The Listener*, September 8, 1955, pp. 370–71.

PARTICLE 9

28 *"vagabond and a loner"*: Einstein, "Autobiographical Sketch," in *Einstein on Einstein*, p. 145.

29 *"Einstein will one day be a great man"*: Carl Seelig, *Albert Einstein: Leben und Werk eines Genies unserer Zeit* (Europa Verlag, 1960), p. 55.

29 *"notes could have been published"*: Einstein to Elizabeth Grossmann, September 20, 1936, AEA 11-481.

29 *"everything else involved subtleties"*: Einstein, "Autobiographical Sketch," in *Einstein on Einstein*, p. 145.

29 *"never bothered about mathematics"*: Seelig, *A Documentary Biography*, p. 28.

30 *"You're a very clever boy"*: Ibid.

31 *"with a divine zeal"*: Einstein, "Autobiographical Sketch," in *Einstein on Einstein*, p. 145.

31 *"absolute space, absolute time"*: Henri Poincaré, *Science and Hypothesis* (Walter Scott, 1905), p. 90.

31 *"reprimand from the director"*: ETH Record and Grade Transcript, *CPAE* 1:28, p. 27.

32 *"even less talent for those subjects"*: Seelig, *Leben und Werk*, p. 65.

32 *"You can't go like that"*: Ibid.

PARTICLE 10

34 *"Besso's strength"*: Einstein to Heinrich Zangger, December 21, 1926, *CPAE* 15:436, p. 414.

34 *"Michele is a terrible schlemiel"*: Einstein to Mileva Marić, March 27, 1901, *CPAE* 1:94, p. 161.

PARTICLE 11

35 *"I would have replied immediately"*: Marić to Einstein, after October 20, 1897, *CPAE* 1:36, p. 34.

36 *"seems to be a very good girl"*: Desanka Trbuhović-Gjurić, *Im Schatten Albert Einsteins: Das tragische Leben der Mileva Einstein-Marić* (Verlag Paul Haupt, 1993), p. 53.

37 *"each other's black souls"*: Einstein to Marić, August 10, 1899, *CPAE* 1:52, p. 131.

37 *"How proud I will be"*: Einstein to Marić, September 13, 1900, *CPAE* 1:75, p. 149.

37 *"Don't be angry with me"*: Einstein to Marić, April 16, 1898, *CPAE* 1:40, p. 124.

37 *"especially the latter"*: Einstein to Marić, March 13 (20), 1897, *CPAE* 1:45, p. 126.

37 *"Have I told you the one"*: Peter Michelmore, *Einstein: Profile of the Man* (Dodd, Mead, 1962), p. 43.

38 *"I would never have the courage"*: Seelig, *A Documentary Biography*, p. 38.

38 *"We shall remain students"*: Einstein to Marić, December 12, 1901, *CPAE* 1:127, p. 186.

38 *"My dear Johnnie"*: Marić to Einstein, 1900, *CPAE* 1:61, p. 138.

PARTICLE 12

39 *"So, what will become,"* and all subsequent quotations: Einstein to Marić, July 29, 1900, *CPAE* 1:68, p. 141.

PARTICLE 13

41 *"I will soon have honored"*: Einstein to Marić, April 4, 1901, *CPAE* 1:96, p. 163.

42 *"Because your work on general chemistry"*: Einstein to Wilhelm Ostwald, March 19, 1901, *CPAE* 1:92, p. 159.

42 *"A few weeks ago"*: Einstein to Ostwald, April 3, 1901, *CPAE* 1:95, p. 162.

PARTICLE 14

43 *"Body height"*: Military Service Book, *CPAE* 1:91, p. 158.

PARTICLE 15

44 *"You absolutely must come"*: Einstein to Marić, April 30, 1901, *CPAE* 1:102, p. 167.

44 *"This evening I sat"*: Einstein to Marić, second half of May 1901, *CPAE* 1:110, p. 173.

44 *"open arms"*: Marić to Helene Savić, second half of May 1901, *CPAE* 1:109, p. 172.

45 *"we rented a very small sleigh"*: Ibid.

45 *"How delightful it was"*: Einstein to Marić, second half of May 1901, *CPAE* 1:107, p. 171.

PARTICLE 16

46 *"this beautiful piece of work"*: Einstein to Marić, May 28, 1901, *CPAE* 1:111, p. 174.

47 *"our little son and your doctoral thesis"*: Ibid.

47 *"turned out to be a Lieserl"*: Einstein to Marić, February 4, 1902, *CPAE* 1:134, p. 191.

PARTICLE 17

49 *"holder of the fed poly. teacher's diploma"*: Advertisement for Private Lessons, *CPAE* 1:135, p. 192.

49 "Herein!": Einstein, *Letters to Solovine*, p. 6.

50 *"I will be glad to see you"*: Ibid., p. 7.

50 *"The menu ordinarily consisted"*: Ibid., p. 8.

51 *"Expert in the noble arts"*: Maurice Solovine, "Dedication of the Olympia Academy, A.D. 1903," *CPAE* 5:3, p. 5.

51 *"Is it as good as all that?"*: Einstein, *Letters to Solovine*, p. 11.

52 *"Say, do you know"*: Ibid.

52 "Amicis carissimis ova dura": Ibid., p. 12.

52 *"We would reach the summit"*: Ibid., p. 14.

PARTICLE 18

55 *"everything the inventor says is wrong"*: Max Flückiger, *Albert Einstein in Bern: Das Ringen um ein neues Weltbild: Eine dokumentarische Darstellung über den Aufstieg eines Genies* (Paul Haupt, 1974), p. 58.

PARTICLE 20

62 *"pulled out from under one"*: Einstein, "Autobiographical Notes," in *Einstein on Einstein*, p. 169.

62 *"transfers its entire energy"*: Einstein, "On a Heuristic Point of View Concerning the Production and Transformation of Light," March 17, 1905, *CPAE* 2:14, p. 100.

PARTICLE 21

67 *"inconsequential babble"*: Einstein to Conrad Habicht, May 18 or 25, 1905, *CPAE* 5:27, p. 19.

67 *"you frozen whale"*: Ibid. p. 20.

PARTICLE 22

71 *"The introduction of a 'light ether'"*: Einstein, "On the Electrodynamics of Moving Bodies," June 30, 1905, *CPAE* 2:23, p. 141.

72 *"seemingly incompatible"*: Ibid., p. 140.

72 *"almost a year fruitlessly"*: Einstein, "How I Created the Theory of Relativity," Kyoto lecture, December 14, 1922, *CPAE* 13:399, p. 637.

72 *"I'm going to give it up"*: Reiser, *Albert Einstein*, p. 68.

72 *"I've completely solved my problem"*: *CPAE* 13:399, p. 637.

72 *"no audible tick-tock"*: Einstein, "The Principal Ideas of the Theory of Relativity," December 1916, *CPAE* 6:44a (in vol. 7, p. 5).

74 *"In conclusion, let me note"*: Einstein, "On the Electrodynamics," p. 171.

PARTICLE 23

75 *"I do not need wine"*: Charles Nordmann, "With Einstein on the Battle Fields," *L'Illustration*, April 15, 1922.

75 *"dead drunk under the table"*: Einstein to Habicht, July 20, 1905–Summer 1915, *CPAE* 5:30, p. 21.

75 *"amusing and seductive"*: Einstein to Habicht, June 30–September 22, 1905, *CPAE* 5.28, p. 21.

76 *"If a body releases the energy"*: Einstein, "Does the Inertia of a Body Depend Upon Its Energy Content?" September 27, 1905, *CPAE* 2:24, p. 174.

PARTICLE 24

78 *"I am doing fine"*: Einstein to Alfred Schnauder, January 5–May 11, 1907, *CPAE* 5:43, p. 28.

78 *"an imposing, impertinent fellow"*: Einstein to Maurice Solovine, April 27, 1906, *CPAE* 5:36, p. 25.

78 *"one of the nicest toys"*: G. J. Whitrow, ed., *Einstein, the Man and His Achievement: A Series of Broadcast Talks* (BBC, 1967), p. 19.

PARTICLE 25

79 *"a modest little band"*: Max Planck to Einstein, July 6, 1907, *CPAE* 5:47, p. 31.

79 *"I must tell you quite frankly"*: Jakob Laub to Einstein, March 1, 1908, *CPAE* 5:91, p. 63.

80 *"I just don't understand"*: Seelig, *A Documentary Biography*, p. 131.

PARTICLE 26

81 *"Oh, that Einstein"*: Constance Reid, *Hilbert* (Springer, 1970), p. 105.

82 *"Gentlemen! The views of space"*: Hermann Minkowski, lecture at University of Cologne, September 21, 1908.

83 *"superfluous learnedness"*: Abraham Pais, *Subtle Is the Lord: The Science and the Life of Albert Einstein* (Oxford University Press, 1982), p. 152.

83 *"I do not understand it"*: Arnold Sommerfeld, quoted in *Albert Einstein: Philosopher-Scientist*, vol. 1, ed. Paul Arthur Schilpp (Library of Living Philosophers, 1949), p. 102.

PARTICLE 27

85 *"size up the beast"*: Einstein to Laub, May 19, 1909, *CPAE* 5:161, p. 120.

85 *"quite unnecessary"*: Reiser, *Albert Einstein*, p. 72.

85 *"Contrary to my habit"*: *CPAE* 5:161, p. 120.

86 *"Israelites among scholars"*: Alfred Kleiner, report to faculty, March 4, 1909, University of Zurich Archive.

86 *"the guild of whores"*: *CPAE* 5:161, p. 120.

PARTICLE 28

87 *"I probably cherish"*: Einstein to Anna Meyer-Schmid, May 12, 1909, *CPAE* 5:154, p. 115.

88 *"It was wrong"*: Einstein to Georg Meyer, June 7, 1909, *CPAE* 5:166, p. 127.

88 *"if I saw the girl again"*: Einstein to Marić, September 28, 1899, *CPAE* 1:57, p. 136.

89 *"I continue to live"*: Einstein to Winteler, September 15, 1909, *CPAE* 5:177a (in vol. 15, p. 16).

89 *"I think of you with heartfelt"*: Einstein to Winteler, March 7, 1910, *CPAE* 5:198a (in vol. 15, pp. 16–17).

90 *"watching my grave"*: Einstein to Winteler, August 7, 1910, *CPAE* 5:218a (in vol. 15, p. 17).

PARTICLE 29

91 *"some silly mathematical transformation"*: Seelig, *A Documentary Biography*, p. 100.

92 *"for fear it might explode"*: Ibid., p. 102.

92 *"a glance"*: David Reichinstein, *Albert Einstein: A Picture of His Life and His Conception of the World* (Edward Goldston, 1934), p. 48.

93 *"See if you can spot"*: Seelig, *A Documentary Biography*, p. 104.

PARTICLE 30

94 *"not people with natural sentiments"*: Einstein to Michele Besso, May 13, 1911, *CPAE* 5:267, p. 187.

95 *"You are this man Kepler"*: Frank, *Einstein*, p. 85.

PARTICLE 31

97 *"A Story of Love"*: *Le Journal*, November 4, 1911.

97 *"A Romance in the Laboratory"*: *Le Petit Journal*, November 5, 1911.

97 *"It has been known"*: Einstein to Zangger, November 7, 1911, *CPAE* 5:303, p. 219.

98 *"boor and a coward"*: *L'Oeuvre*, November 23, 1911.

98   *"Do not laugh at me"*: Einstein to Marie Curie, November 23, 1911, *CPAE* 5:312a (in vol. 8, p. 6).

PARTICLE 32

100 *"happiest thought"*: Einstein, "Fundamental Ideas and Methods of the Theory of Relativity, Presented in Their Development," 1920, unpublished draft of a paper for *Nature*, *CPAE* 7:31, p. 136. It can also be translated as "luckiest thought."

101 *"I decided to"*: *CPAE* 13:399, p. 638 of original volume.

104 *"Grossmann, you've got to help"*: Louis Kollros, *Helvetica Physica Acta Supplement* 4 (1956), p. 271.

107 *"Nature shows us only the tail"*: Einstein to Zangger, March 10, 1914, *CPAE* 5:513, p. 381.

PARTICLE 33

109 *"I can't even begin"*: Einstein to Elsa Löwenthal, April 30, 1912, *CPAE* 5:389, p. 291.

109 *"I am writing so late"*: Einstein to Löwenthal, May 21, 1912, *CPAE* 5:399, p. 300.

109 *"afraid of the relatives"*: Einstein to Löwenthal, August 11, 1913, *CPAE* 5:465, p. 348.

110 *"I treat my wife as an employee"*: Einstein to Löwenthal, December 2, 1913, *CPAE* 5:488, p. 365.

110 *"It is almost disgraceful"*: Einstein to Löwenthal, October 16, 1913, *CPAE* 5:478, p. 357.

110 *"bristly girlfriend"*: Einstein to Löwenthal, October 10, 1913, *CPAE* 5:476, p. 356.

110 *"The hairbrush is being"*: Einstein to Löwenthal, November 22, 1913, *CPAE* 5:486, p. 363.

110 *"But if I were to start"*: Einstein to Löwenthal, after December 2, 1913, *CPAE* 5:489, p. 366.

PARTICLE 34

111 *"I have firmly decided"*: Einstein to Löwenthal, after August 11, 1913, *CPAE* 5:466, p. 348.

111 *"somewhat calm and objective judgment"*: Einstein, to Montreal Pipe Smokers Club, quoted in *New York Times*, March 12, 1950, AEA 60-125.

PARTICLE 35

113 *"A. You will make sure"*: Einstein, "Memorandum to Mileva Einstein-Marić, with Comments," July 18, 1914, *CPAE* 8:22, pp. 32–33.

114 *"completely clear about the situation"*: Einstein to Marić, ca. July 18, 1914, *CPAE* 8:23, p. 33.

115 *"I would be a real monster"*: Einstein to Löwenthal, July 26, 1914, *CPAE* 8:27, p. 36.

PARTICLE 36

116 *"less impressed by Lenin than I expected to be"* and all subsequent quotations: Bertrand Russell, interview with John Chandos, April 11 and 12, 1961.

PARTICLE 37

117 *"saber rattling"*: Einstein to Helene Savić, December 17, 1912, *CPAE* 5:424, p. 325.

118 *"At such a time as this"*: Einstein to Paul Ehrenfest, August 19, 1914, *CPAE* 8:34, p. 41.

118 *"aggressive characteristics"*: Einstein, "My Opinion on the War," October 23–November 11, 1915, *CPAE* 6:20, pp. 96–97.

PARTICLE 38

122 *"I was beside myself"*: Einstein to Ehrenfest, January 17, 1916, *CPAE* 8:182, p. 179.

122 *"the strongest emotional experience"*: Pais, *Subtle Is the Lord*, p. 253.

PARTICLE 39

124 Charles Chaplin, *My Autobiography* (Simon & Schuster, 1964), pp. 320–21.

PARTICLE 40

127 *"He is one of the most excellent"*: Einstein to Kathia Adler, February 20, 1917, AEA 43-3.

127 *"the Objectivity reflected"*: "Friedrich Adler als Physiker. Eine Unterredung mit A. Einstein," *Vossische Zeitung*, May 23, 1917, morning edition, p. 2.

PARTICLE 41

128 *"The war is kindly disposed"*: Karl Schwarzschild to Einstein, December 22, 1915, *CPAE* 8:169, p. 164.

PARTICLE 42

131 *"greatest mistake"*: In fact, in the quotation from which this springs, Einstein called the cosmological constant his "biggest blunder": George Gamow, *My World Line: An Informal Autobiography* (Viking, 1970), p. 44.

133 *"a slight modification"*: Einstein, "Cosmological Considerations in the General Theory of Relativity," February 15, 1917, *CPAE* 6:43, p. 424.

133 *"In order to arrive at"*: Ibid., p. 432.

134 *"gravely detrimental"*: Einstein, "Do Gravitational Fields Play an Essential Part in the Structure of the Elementary Particles of Matter?," April 24, 1919, *CPAE* 7:17, p. 83.

134 *"Nature is the realization of the"*: Einstein, "On the Method of Theoretical Physics," AEA 96-38, in *Ideas and Opinions* (Bonanza, 1954), p. 274.

136 *"Well, my husband"*: Bennett Cerf, *Try and Stop Me: A Collection of Anecdotes and Stories, Mostly Humorous* (Simon & Schuster, 1944), p. 163.

PARTICLE 43

139 *"Yesterday, the question was suddenly raised"*: Ilse Einstein to Georg Nicolai, May 22, 1918, *CPAE* 8:545, p. 564.

PARTICLE 44

141 *"overcome anti-Semitism by dropping"*: Einstein, "Assimilation and Anti-Semitism," April 3, 1920, *CPAE* 7:34, p. 154.

141 *"pussyfooting"*: Einstein, "A Confession," April 5, 1920, *CPAE* 7:37, p. 159.

141 *"I felt a duty to oppose"*: Einstein to Julius Katzenstein, December 27, 1931, AEA 78-936.

142 *"community of tradition"*: Albert Einstein, "Why Do They Hate the Jews," *Collier's*, November 26, 1938.

142 *"I am* against *nationalism"*: Kurt Blumenfeld, "Einstein and Zionism," in Seelig, *A Documentary Biography*, p. 74.

142 *"a center of culture"*: Einstein, "Jewish Recovery," AEA 28-164, in *Ideas and Opinions*, p. 184.

143 *"most amusing"*: Clark, *Einstein*, p. 318.

PARTICLE 46

149 *"I do not see why"*: Subrahmanyan Chandrasekhar, "Verifying the Theory of Relativity," *Bulletin of Atomic Sciences* 31:6 (June 1975), pp. 17–22.

149 *"REVOLUTION IN SCIENCE"*: London *Times*, November 7, 1919.

150 *"After a careful study of the plates"*: Ibid.

150 *"We owe it to that great man"*: Ibid.

150 *"since Newton's day"*: Ibid.

150 *"the famous physicist"*: Ibid.

151 *"ECLIPSE SHOWED GRAVITY VARIATION"*: *New York Times*, November 9, 1919.

151 *"LIGHTS ALL ASKEW"*: *New York Times*, November 10, 1919.

152 *"You must be one of"*: Subrahmanyan Chandrasekhar, *Eddington: The Most Distinguished Astrophysicist of His Time* (Cambridge University Press, 1983), p. 30.

152 *"a man dropping"*: *New York Times*, December 3, 1919.

153 *"Here [in Berlin] all conditions are variable"*: Einstein to Zangger, December 15, 1919, *CPAE* 9:217, p. 186.

153 *"A new luminary"*: *Berliner Illustrirte Zeitung*, December 14, 1919.

PARTICLE 47

154 *"shown around like a prize"*: Einstein to Besso, before May 30, 1921, *CPAE* 12:141, p. 103.

155 *"All my life I have"*: Zelda Popkin, *Open Every Door* (Dutton, 1956), p. 136.

155 *"During the crossing"*: Seelig, *A Documentary Biography*, p. 136.

155 *"Your leader, Dr. Weizmann"*: *New York Times*, April 13, 1921.

155 *"a new theory of eternity"*: Interview of Harlow Shapley by Charles Weiner and Helen Wright on June 8, 1966, Niels Bohr Library & Archives, American Institute of Physics, College Park, MD, USA, www.aip .org/history-programs/niels-bohr-library/oral-histories/4888-1.

156 *"It's a miracle"*: Einstein to Besso, before May 30, 1921, *CPAE* 12:141, p. 103.

PARTICLE 48

158 *"the highest form of musicality"*: Einstein, "Autobiographical Notes," in *Einstein on Einstein*, p. 169.

159 *"Don't frown like that!"*: Einstein to Niels Bohr, March 2, 1955, AEA 33-204.

159 *"so incredibly sweet"*: Niels Bohr interviewed in Tisvilde by Aage Bohr and Léon Rosenfeld, July 12, 1961, Niels Bohr Archive, Copenhagen.

PARTICLE 49

161 *"Fundamental Ideas and Problems of the Theory of Relativity"*: Einstein, lecture delivered to the Assembly of Nordic Naturalists in Gothenburg on July 11, 1923, *CPAE* 14:75, pp. 74–81.

PARTICLE 50

162 *"We drove in individual little"*: Einstein, "Travel Diary: Japan, Palestine, Spain," October 28, pp. 6v–7v; *CPAE* 13:379, p. 301.

163 *"Industrious, dirty, numbed"*: "Travel Diary," November 10, p. 13; *CPAE* 13:379, p. 305.

163 *"most pitiful of people"*: "Travel Diary," January 5, p. 29v; *CPAE* 13:379, p. 318.

163 *"similar to Italians"*: "Travel Diary," December 5, p. 22v; *CPAE* 13:379, p. 312.

163 *"Earnest respect without"*: "Travel Diary," December 10, p. 24v; *CPAE* 13:379, p. 314.

163 *"intellectual demands"*: "Travel Diary," December 5, p. 23; *CPAE* 13:379, pp. 312–13.

164 *"dull ethnic brethren"*: "Travel Diary," February 3, p. 34; *CPAE* 13:379, pp. 321–22.

PARTICLE 51

165 *"A calm and modest life"*: Einstein to a courier, 1922, AEA 124-552.

165 *"Where there's a will"*: Einstein to a courier, 1922, AEA 124-553.

PARTICLE 52

167 *"a cross between"*: Frank, *Einstein*, p. 191.

167 *"Organic"*: Arnold Whittick, *Eric Mendelsohn* (Faber & Faber, 1940), p. 64.

PARTICLE 53

168 Chapter draws on Esther Salaman, "A Talk with Einstein," *The Listener*, 54:1384, September 8, 1955, pp. 370–71.

PARTICLE 54

171 *"You have a father"*: Einstein to Hans Albert Einstein, October 13, 1916, *CPAE* 8:263, pp. 252–53.

172 *"It would be a crime"*: Einstein to Marić, December 23, 1925, *CPAE* 15:135, p. 156.

172 *"She was the first"*: Einstein to Hans Albert Einstein, February 23, 1927, *CPAE* 15:484, p. 483.

172 *"good-looking woman"*: Einstein to Marić, October 17, 1925, *CPAE* 15:88, p. 101.

172 *"If you ever feel the need"*: *CPAE* 15:484, p. 483.

173 *"I don't understand it"*: Michelmore, *Einstein*, p. 131.

PARTICLE 55

174 *"I am a little girl of six"*: Ann to Einstein, 1951, AEA 42-653, in *Dear Professor Einstein: Albert Einstein's Letters to and from Children*, ed. Alice Calaprice (Prometheus, 2002), p. 188.

174 *"I have a problem"*: Anna Louise (Falls Church, Virginia) to Einstein, February 8, 1950, AEA 42-642, in ibid., p. 175.

175 *"My mother said"*: Frank (Bristol, Pennsylvania) to Einstein, March 25, 1950, AEA 42-644, in ibid., p. 178.

175 *"We would like to know"*: Kenneth (Asheboro, North Carolina) to Einstein, August 19, 1947, AEA 42-618.1, in ibid., p. 144.

175 *"tell me what Time is"*: Peter (Chelsea, Massachusetts) to Einstein, March 13, 1947, AEA 616.1, in ibid., p. 141.

175 *"my father and I"*: John (Culver, Indiana) to Einstein, 1952, AEA 42-663, in ibid., p. 193.

175 *"to find out if you really exist"*: June (British Columbia, Canada) to Einstein, June 3, 1952, AEA 42-662, in ibid., p. 197.

175 *"not aware that you were still alive"*: Myfanwy (South Africa) to Einstein, July 10, 1946, AEA 42-611, in ibid., p. 149.

176 *"There* will *be a remedy"*: Einstein to Myfanwy (South Africa), August 25, 1946, AEA 42-612, in ibid., p. 153.

176 *"Your gift will be"*: Einstein to the fifth-grade class of Farmingdale Elementary School, March 26, 1955, AEA 42-722, in ibid., p. 219.

PARTICLE 56

178 *"God does not play dice"*: Einstein's wording was often different from this, though he used several versions of the same metaphor. For examples, see Einstein to Max Born, December 4, 1926, *CPAE* 15:426, p. 403; Niels Bohr, in Schilpp, *Philosopher-Scientist*, vol. 1, p. 218; Einstein to Born, September 7, 1944, AEA 8-207.

178 *"It cannot be for us"*: Werner Heisenberg, *Encounters with Einstein: And Other Essays on People, Places, and Particles* (Princeton University Press, 1989), p. 117.

PARTICLE 57

180 *"a big quantum egg"*: Einstein to Ehrenfest, November 20, 1925, *CPAE* 15:114, p. 136.
180 *"We cannot observe electron"*: Werner Heisenberg, *Physics and Beyond: Encounters and Conversations* (Harper & Row, 1971), p. 63.
181 *"A new fashion"*: Frank, *Einstein*, p. 216.

PARTICLE 58

182 *"Einstein sublimely dignified"* and all other quotations: Harry Kessler, *Berlin in Lights: The Diaries of Count Harry Kessler, 1918–1937* (Weidenfeld & Nicolson, 1971), p. 281.

PARTICLE 59

185 *"One can't make a theory"*: Wolfgang Pauli, *Writings on Physics and Philosophy* (Springer, 1994), p. 121.
185 *"Every night, Bohr came to"*: Paul Ehrenfest to S. A. Goudsmit, G. E. Uhlenbeck, and G. H. Dieke, November 3, 1927, in *Niels Bohr Collected Works*, vol. 6: *Foundations of Quantum Physics I (1926–1932)*, ed. Jørgen Kalckar (North Holland, 1985), pp. 415–18.
186 *"Einstein, a majestic figure"*: Léon Rosenfeld, "Some Concluding Remarks and Reminiscences," in *Proceedings, 14th Solvay Conference on Physics: Fundamental Problems in Elementary Particle Physics: Brussels, Belgium, October, 1967* (Interscience, 1968), p. 232.

PARTICLE 60

188 *"Here lies an old corpse"*: *New York Times*, March 27, 1972.
189 *"sitting with a pretty girl"*: Jamie Sayen, *Einstein in America: The Scientist's Conscience in the Age of Hitler and Hiroshima* (Crown, 1985), p. 130.
189 *"That is Miss Dukas"*: Michael Grüning, *Ein Haus für Albert Einstein: Erinnerungen, Briefe, Dokumente* (Verlag der Nation, 1990), p. 51.

PARTICLE 61

192   *"Life is short"*: The letter he wrote is lost. Reported in *Berliner Tageblatt*, May 14, 1929, and recollected by the architect Konrad Wachsmann; see Grüning, *Ein Haus*, p. 122ff.

PARTICLE 62

193   *"Do you believe in God"*: "Einstein Believes in Spinoza's God," *New York Times*, April 24, 1929; also, Einstein to Herbert S. Goldstein, April 25, 1929, AEA 33-272.

194   *"Neither intellect nor"*: Benedict de Spinoza, *The Ethics*, Part I, Proposition XVII, Note.

194   *"I cannot conceive"*: Albert Einstein, "What I Believe," *Forum and Century* 84:4, October 1930, AEA 78-645.

194   *"Whatsoever is, is in God"*: Spinoza, *Ethics*, Part I, Prop. XXIX, Proof.

194   *"We followers of Spinoza"*: Einstein to Eduard Büsching, October 25, 1929, AEA 33-275.

194   *"I do not believe in free will"*: Albert Einstein, "My Credo," to the German League of Human Rights, Berlin, Autumn 1932, AEA 28-218.

195   *"We are in the position"*: George Sylvester Viereck, *Glimpses of the Great* (Duckworth, 1930), p. 373.

PARTICLE 63

197   *"Now, but only now"*: *Daily Chronicle*, January 26, 1929.

197   *"So, you were right"*: Einstein to Wolfgang Pauli, January 22, 1932, AEA 19-169.

197   *"I shall never ever"*: Einstein to Solovine, November 25, 1948, AEA 21-256.

PARTICLE 64

198   *"a very small star"*: Arnold J. Toynbee, *Acquaintances* (Oxford University Press, 1967), p. 268.

199   *"Your dismay towards"*: Einstein to Elsa Einstein, May 1931; part of sealed correspondence released in 2006, AEA 143-242.

199 *"Her chasing me"*: Einstein to Margot Einstein, May 1931; part of sealed correspondence released in 2006, AEA 143-292.

199 *"The small package really"*: Einstein to Ethel Michanowski, May 24, 1931, AEA 84-104.

200 *"You have to see him"*: Elsa Einstein to Hermann Struck and wife, 1929, Archives of the Max Planck Society, Va. Abt., Rep. 2, Nr. 50.

PARTICLE 65

201 *Spring was slowly becoming summer*: William Golding, "Thinking as a Hobby," *Holiday*, August 1961, pp. 8, 10–13.

PARTICLE 66

203 *"Why do you emphasize"*: Einstein to Sigmund Freud, March 22, 1929, *CPAE* 16:465, p. 416.

203 *"Is there any way of delivering mankind"*: All quotations from Einstein's letter: Einstein to Sigmund Freud, July 30, 1932, AEA 32-543.

205 *"You are amazed that it is so easy"*: All quotations from Freud's letter: Freud to Einstein, September 1–8, 1932, AEA 32-548.

PARTICLE 67

207 *"Albert Einstein believes in, advises"*: Woman Patriot Corporation memo to the US State Department, November 22, 1932, contained in Einstein's FBI file, Section 1, available at https://vault.fbi.gov/Albert%20Einstein.

208 *"What is your political creed?"*: See *New York Times*, December 6, 1932, pp. 1, 18.

PARTICLE 68

211 *"I think and concentrate"*: Mary Pickford, *Sunshine and Shadows* (Doubleday, 1955), pp. 230–31.

212 *"Hallo! Why don't you"*: Chaplin, *Autobiography*, pp. 322–23.

PARTICLE 69

213 *"the empty stomach of Germany"*: New York Times, December 12, 1930.

214 *"If I were as you want"*: Elsa Einstein to Antonina Vallentin, Caputh, June 6, 1932, AEA 79-212.

215 *"Dependence on the Prussian government"*: Einstein to Prussian Academy, March 28, 1933, AEA 36-55.

215 *"Even though on political"*: Max Planck to Heinrich von Ficker, March 31, 1933, Prussian Secret State Archives, GStA PK, I. HA Rep. 76 Kultusministerium, Vc Sekt. 2 Tit. XXIII Litt. F Nr. 2 Bd. 16. *Appointment and salary of the members of the Academy of Sciences in Berlin, Vol. 16, 1933–1934.*

216 *"In spite of everything"*: Einstein to Planck, April 6, 1933, AEA 19-392.

PARTICLE 70

218 *"The sorrow is eating Albert"*: Vallentin, *Drama of Albert Einstein*, p. 196.

218 *"about which nothing"*: Einstein to Besso, October 21, 1932, AEA 7-370.

219 *"You have probably already"*: Einstein to Carl Seelig, January 4, 1954, AEA 39-59.

PARTICLE 71

220 *"one of those people"*: Lady Margaret Proby to Oliver Locker-Lampson, October 25, 1914, Norfolk Records Office, Norwich, Uncatalogued Box/2190/1.

221 *"legendary hero"*: Daily Mirror, September 30, 1930.

222 *"He is an eminently"*: Einstein to Elsa Einstein, July 21, 1933, AEA 143-250.

222 *"turned out her most glorious citizen"*: Oliver Locker-Lampson, speech to the House of Commons, July 26, 1933.

PARTICLE 72

223 *"The husband of the second"*: Otto Nathan and Heinz Norden, ed., *Einstein on Peace* (Schocken, 1960), p. 227.

224 *"What I have to say will"*: Einstein to Alfred Nahon, July 20, 1933, AEA 51-227.

PARTICLE 73

226 *"I didn't know it was"*: Manchester *Guardian*, September 8, 1933.

226 *"when a bandit"*: *New York Times*, September 9, 1933.

227 *"If any unauthorized person"*: *New York Times*, September 11, 1933.

227 *"The beauty of my guards"*: Dimitri Marianoff, with Palma Wayne, *Einstein: An Intimate Study of a Great Man* (Doubleday, Doran, 1944), p. 161.

PARTICLE 74

228 Epstein's account of the Einstein sitting: Jacob Epstein, *Let There Be Sculpture: An Autobiography* (Michael Joseph, 1940), pp. 94–96.

PARTICLE 75

230 *"quaint and ceremonious"*: Einstein to Queen Elisabeth of Belgium, November 20, 1933, AEA 32-369.

231 *"Please don't tell anybody"*: Churchill Eisenhart, "Albert Einstein, As I Remember Him," *Journal of the Washington Academy of Sciences* 54:8 (1964), p. 325.

PARTICLE 76

232 *"Physics should represent"*: Einstein to Max Born, March 3, 1947, in *The Born–Einstein Letters: Correspondence Between Albert Einstein and Max and Hedwig Born from 1916–1955, with Commentaries by Max Born* (Macmillan, 1971), p. 158.

PARTICLE 77

236 *"And if my Ilse"*: Vallentin, *Drama of Albert Einstein*, p. 238.

236 *"He wanders around"*: Ibid.

236 *"Oh, I shall really miss her"*: Peter A. Bucky, *The Private Albert Einstein* (Andrews and McMeel, 1992), p. 13.

236 *"But as long as I am able"*: Einstein to Hans Albert Einstein, January 4, 1937, AEA 75-926.

PARTICLE 78

237　*"At close quarters"*: C. P. Snow, "On Albert Einstein," *Commentary*, March 1967.

PARTICLE 79

240　*"The dog is very smart"*: Anita Ehlers, *Liebes Hertz!: Physiker und Mathematiker in Anekdoten* (Springer-Verlag, 1994), p. 162.

240　*"I know what's wrong"*: Banesh Hoffmann and Helen Dukas, *Albert Einstein: Creator and Rebel* (Viking, 1972), p. 252.

240　*"Let's walk quickly"*: Ibid.

PARTICLE 80

241　*"My trip to this university"*: *Baltimore Afro-American*, May 11, 1946.

242　*"complete artistic mastery"*: *Daily Princetonian*, April 16, 1937.

242　*"would have been offended"*: Marian Anderson, *My Lord, What a Morning* (Viking, 1956), p. 267.

PARTICLE 81

244　All quotations are from Einstein's FBI file, Section 1.

PARTICLE 82

246　Jerome Weidman, "The Night I Met Einstein," *Reader's Digest*, November 1955.

PARTICLE 83

250　*"is so simple God could"*: Hoffman and Dukas, *Einstein*, p. 228.

PARTICLE 84

253 *"Sundials"*: Joan Rothman Brill, *My Father and Albert Einstein: Biography of a Department Store Owner, Whose Thirst for Knowledge Enabled His Close Friendship with a Genius Who Changed Man's Concepts of the Universe* (iUniverse, 2008), p. 6.

PARTICLE 85

255 *"Alex, what are you up to?"*: Ruth Moore, *Niels Bohr: The Man, His Science, and the World They Changed* (Knopf, 1966), p. 268.
256 *"my duty to bring to your attention"*: Einstein to Franklin Roosevelt, August 2, 1939, AEA 33-143.
257 *"I understand that Germany"*: Ibid.
257 *"Alex, what you are after"*: Alexander Sachs, before Senate special committee on atomic energy hearing, November 27, 1945.

PARTICLE 86

258 *"savant and genius"*: 73rd Congress, 2nd Session, House Joint Resolution 309, March 28, 1934, AEA 50-673.
259 *"a way of life tied"*: "Einstein Is Sworn as Citizen of U.S.," *New York Times*, October 2, 1940.

PARTICLE 87

260 *"Dear Einstein"*: Born to Einstein, July 15, 1944, AEA 8-206.
261 *"Dear Born"*: Einstein to Born, September 7, 1944, AEA 8-207.

PARTICLE 88

263 *"Once I am set"*: Ernst G. Straus, "Memoir," in *Einstein: A Centenary Volume*, ed. A. P. French (Harvard University Press, 1979), p. 31.

PARTICLE 89

264 *Bohr visited the Institute*: Abraham Pais, *Niels Bohr's Times: In Physics, Philosophy, and Polity* (Oxford University Press, 1991), pp. 12–13.

PARTICLE 90

267 *"This kind of inquisition"*: Einstein to William Frauenglass, May 16, 1953, AEA 41-112.

267 *"enemy of America"*: *New York Times*, June 14, 1953.

268 *"I would rather choose"*: Einstein to *Reporter Magazine*, November 1954, AEA 90-187.

268 *"Since my ambition has always"*: Stanley Murray to Einstein,November 11, 1954, AEA 41-858.

PARTICLE 91

271 *attend his American citizenship hearing*: Oskar Morgenstern, "Account of Kurt Gödel's naturalization," September 13, 1971, Dorothy Morgenstern Thomas collection related to Kurt Gödel, from the Shelby White and Leon Levy Archives Center, Institute for Advanced Study, Princeton, NJ, USA.

PARTICLE 92

273 *"the greatest Jew alive"*: "Israel: Einstein Declines," *Time*, December 1, 1952.

273 *"If I were to be president"*: Sayen, *Einstein in America*, p. 247.

274 *"I am not the person"*: Ibid., p. 246.

274 *"Acceptance would entail"*: Abba Eban to Einstein, November 17, 1952, AEA 41-84.

274 *"I am deeply moved"*: Einstein to Eban, November 18, 1952, AEA 28-943.

275 *"Tell me what to do"*: Navon, in Holton and Elkana, *Historical and Cultural Perspectives*, p. 295.

275 *"much rather see reasonable agreement"*: Einstein, from an address delivered at the Commodore Hotel, New York, April 17, 1938, published in *New Palestine*, April 29, 1938, AEA 28-427.

PARTICLE 93

276 The letter quoted in this chapter is from Einstein to Vero and Bice Besso,
     March 21, 1955, AEA 7-245.

PARTICLE 94

278 *"The strange thing about"*: Einstein to Queen Mother Elisabeth of Bel-
     gium, January 12, 1953, AEA 32-405.
279 *"It is tasteless"*: Helen Dukas, "Einstein's Last Days," April 1955, AEA
     39-71.

PARTICLE 97

286 *"It's a boy"*: Edward Teller, with Judith L. Shoolery, *Memoirs: A Twentieth-
     century Journey in Science and Politics* (Perseus, 2001), p. 352.
288 *"the father of the release"*: *Atlantic Monthly*, November 1945.
288 *"annihilation of all life"*: *Today with Mrs. Roosevelt*, February 12, 1950,
     AEA 96-318, in Nathan and Norden, *Einstein on Peace*, p. 521.

PARTICLE 98

292 *"It seems doubtful altogether"*: Einstein, "Autobiographical Sketch," in Gut-
     freund and Renn, *Einstein on Einstein*, p. 148.
292 *"The fairest thing"*: Einstein, "What I Believe," AEA 78-645, in *The World
     as I See It* (Philosophical Library, 1949), p. 5.

PARTICLE 99

293 *"Of what is significant"*: Georges Schreiber, *Portraits and Self-Portraits*
     (Houghton Mifflin, 1936).

INDEX

ABOUT THE AUTHOR

Samuel Graydon is the science editor at the *Times Literary Supplement*. He has published short fiction and has been longlisted for an Alpine Fellowship. He studied English Language and Literature at the University of Oxford. He and his family live in Bath, UK.